Make: Arduino Bots and Gadgets

Learning by Discovery

Kimmo and Tero Karvinen

with photographs and illustrations by the authors

O'REILLY®

Beijing • Cambridge • Farnham • Köln • Sebastopol • Tokyo

Make: Arduino Bots and Gadgets

by Kimmo and Tero Karvinen

Copyright © 2011 O'Reilly Media, Inc. All rights reserved.
Printed in Canada.

Published by O'Reilly Media, Inc., 1005 Gravenstein Highway North, Sebastopol, CA 95472.

O'Reilly Media books may be purchased for educational, business, or sales promotional use. Online editions are also available for most titles (*my.safaribooksonline.com*). For more information, contact our corporate/institutional sales department: 800-998-9938 or *corporate@oreilly.com*.

Development Editors: Brian Jepson and Brian Sawyer

Production Editor: Holly Bauer

Technical Editor: Joe Saavedra

Copyeditor: Rachel Monaghan

Proofreader: Jennifer Knight

Translator: Marko Tandefelt

Indexer: Ellen Troutman Zaig

Cover Designer: Mark Paglietti

Interior Designer: Ron Bilodeau

Illustrator/Photographer: Kimmo Karvinen

Cover Photographer: Kimmo Karvinen

Software Architect: Tero Karvinen

Print History:

March 2011:   First Edition.

 This book uses Otabind™, a durable and flexible lay-flat binding.

ISBN: 978-1-449-38971-0

[TI]

Contents

Preface

In the early days, embedded systems were built primarily by engineers in a pretty exclusive club. Embedded devices and software tools were expensive, and building a functional prototype required significant software engineering and electrical engineering experience.

With the arrival of Arduino, the open source electronics prototyping platform, things are cheaper and easier. The hardware is inexpensive (around $30), the software is free, and the Arduino environment is designed for artists, designers, and hobbyists rather than engineering professionals.

The ultimate goal of this book is to teach you how to build prototypes using Arduino. We'll offer just enough theory to help you apply your new skills to your own projects. You will also become familiar with the logic behind coding and components. We will explain every single line of code and tell you how each component is used. You will learn by completing actual projects, and the knowledge you gain will enable you to further develop your own ideas.

Most books on embedded systems are either so specialized that you need to work within the particular field or too simplistic to be interesting. Books for beginners often just teach you to blindly follow instructions; here, we aim to promote a deeper understanding and a skill set that can be applied more flexibly.

Finally, this book is meant for readers who want to learn how to build prototypes of interesting gadgets, not for those who want to build a dental X-ray machine or a microwave oven. At the same time, you will be able to apply the techniques covered in the book to make prototypes of commercial device concepts.

Embedded Systems Are Everywhere

An *embedded system* is a microcontroller-based device designed for a very specific purpose. Some examples include washing machines, cell phones, elevators, car brakes, GPS devices, air conditioning units, microwave ovens, wristwatches, and robotic vacuum cleaners. Unlike the user interface you're accustomed to with traditional computers, embedded systems typically do not include a display, mouse, and keyboard. Instead, you might control them via switches and foot pedals, for example.

Most embedded systems are reactive systems, operating in a continuous interaction with their environment and responding within a tempo defined by that environment. This makes them a logical choice for tasks that must react immediately, such as a car braking system.

In some cases, it can be hard to tell whether a particular system should be classified as an embedded system or a computer. For example, cell phones are starting to include more and more features typically associated with computers, but they still have much in common with embedded systems.

Why Should You Study Embedded Systems?

The world is already full of embedded systems. With reasonable effort, you can learn how to build one yourself. Turn inventions and ideas into inexpensive prototypes, automate your home by creating a fish-feeding device or controlling lighting from your computer, or build a remote-controlled surveillance camera for your yard that you can access via a computer located anywhere in the world. Artists can create interactive installations or integrate sensors into a game that you can control without touching a computer. Possible implementations are endless.

During the 2000s, the DIY meme gathered more and more popularity, as is evident with the growth of *MAKE Magazine* and websites such as *http://www.instructables.com*. The Bay Area Maker Faire, an annual DIY festival, went from 22,000 attendees in its first year (2006) to more than double that amount (45,000) in its second year. And each year, Maker Faire attendance keeps growing.

Learning embedded systems is becoming even more appealing due to the growing interest in robotics. In a 2006 *Scientific American* article,* Microsoft founder Bill Gates predicted that robotics would be the next revolution within homes, comparing the current state of the robotics industry to the computing industry in the 1970s. Gates anticipates that robots will soon become a natural part of a home, taking care of simple tasks such as vacuum cleaning, lawn mowing, surveillance, and food service. In addition, because robots can be controlled remotely from anywhere, we'll be able to use them for telepresence—viewing, hearing, and touching people and things without even having to be present.

Intelligent Air Conditioning

The common use of embedded systems is not just the stuff of science fiction or future technology. It's already here and pervasive in the home. Consider air conditioning. A smart air conditioning system adjusts itself based on measurements. How does it know when the air is thick or stale?

Air conditioners measure the temperature, humidity, and sometimes also carbon dioxide levels using sensors. A microcontroller (a small, dedicated computer) follows these measurements, and if the air is damp, for example, it activates a servo that opens an air valve, letting fresh air flow in. This type of

---

http://www.scientificamerican.com/article.cfm?id=a-robot-in-every-home

intelligent air control system has many benefits. It saves energy, because the air conditioning system doesn't need to be used at full power all the time, and it makes working in such a space more comfortable, because there's neither a constant draft nor stagnant air. The heating and air conditioning system at your own school or job likely functions on the same principles.

Sensors, Microcontrollers, and Outputs

Embedded systems include sensors, microcontrollers, and outputs. *Sensors* measure conditions within a physical environment, such as distance, acceleration, light, pressure, reflection of a surface, and motion.

The *microcontroller* is the brain of an embedded system. It's a tiny computer, with a processor and memory, which means you can run your own programs on it. The Arduino microcontroller used in this book is programmed using a full-size computer via a USB cable, with sensors and outputs connected to the microcontroller pins.

Outputs affect the physical environment. Examples of outputs you'll learn to control in this book include LEDs and servo motors. Output devices are sometimes known as *actuators*.

Learn Embedded Systems in a Week

This book will teach you the basics of embedded systems in just one week, during which time you'll build your first gadget. After that, you can move on to more complex projects and prototypes based on your own ideas. Within seven days, you will already be deep within the world of embedded systems.

This goal can sound immense—at least, *we* felt it was impossible before we became familiar with contemporary development environments. But today, many projects that once felt impossible now seem straightforward.

The purpose of this book is to teach you how to build embedded systems, and we've left out any topic that does not support the practice of building prototypes. For example, we don't cover history, movement of electrons, or complex electrical formulas. We believe it makes more sense to study these concepts after you are surrounded by your own homemade devices.

Classroom Use

We tested this book with actual students during a one-week, intensive course led by Tero Karvinen. By the end of the week, all the students in the course were able to build their own prototypes.

The students built many types of projects: a burglar alarm that can be disarmed with a wireless RFID keychain; a flower-measurement device that saves the height, humidity, and temperature of a flower to memory; a sonar device that draws an image of its distance on a computer screen; an automatic triggering device for a camera; a web-based control device for a camera; and a temperature meter observable via an Internet interface. For more examples of projects, visit *http://BotBook.com/*.

Feedback from the class included one common wish: a longer course with more theory. Hopefully, you will become equally hungry for more after you have learned how to build gadgets. We believe that learning electronic theory becomes more interesting after you have already built functional devices. For a complete book on electronics that begins at the beginning, see Charles Platt's *Make: Electronics* (O'Reilly, *http://oreilly.com/catalog/9780596153748*).

What You Need to Know

Being able to use a computer is a prerequisite for completing the exercises in this book. You will need to know how to install programs and solve simple problems that often pop up during program and driver installation.

We've tested the instructions in this book in Ubuntu Linux, Windows 7, and Mac OS X. You should be able to implement the instructions relatively easily for other Windows systems or other Linux distributions.

Programming skills can be helpful but are not necessary for learning embedded systems. The particular programming language you know isn't important, but being familiar with basic programming principles such as functions, if-then statements, loops, and comparisons is beneficial. It's possible to learn programming along with learning about embedded systems, but this approach could take more time. You might find it useful to consult a beginner's book on programming.

High school–level electrical theory and knowledge of voltage, current, resistance, and circuits is sufficient. Have you already forgotten this? No worries— we will revisit basic electrical theory before starting the projects.

How to Read This Book

One of our goals is to provide information in an easily digestible form. By reading this book, anyone can learn how to build impressive-looking electronic devices. Instead of splitting the book into separate sections for techniques and code, we have attempted to combine the information within six projects. This way, you will learn new things bit by bit and can immediately test them in real situations.

The beginning of each project provides learning goals and a list of necessary parts. Before building a device, you can test each part individually; applying the components usually becomes much easier once you understand their core functions. It is useful to come back to these introductory sections later, as you incorporate things you have learned into your own new applications.

We also explain each line of code. This does not mean that you should first read the explanations and continue only after you have internalized everything. We always provide the entire functional code, which you can type or download from *http://BotBook.com/*. Once you have succeeded in getting one version of the code to work, you'll be motivated to find out *how* it works or to customize it for your own purposes. When you start to build your own devices, the explanations will make it easier for you to identify the necessary sections of the provided code.

The projects are partitioned so you can test each part one step at a time. This way, it is easier to understand the function of each step and the relationships between different parts. This also helps ensure that once you have built a device, you can easily troubleshoot any problems; if something doesn't work, you can always go back to an earlier functioning phase and restart from there.

There are examples of enclosures for several projects in this book. They are useful as teaching techniques for mechanical construction and give you ideas for how to make a demonstrable prototype relatively inexpensively. You are not obligated to follow the instructions literally. You might have different parts or a better vision for the look of your device.

Contents of This Book

This book includes two introductory chapters followed by six chapters with projects. As you move through the book, you'll go from learning the basics of Arduino to completing projects with moving parts, wireless communication, and more:

Chapter 1, Introduction

> This chapter explains prototyping, including an overview of the philosophy behind it, techniques, and tools.

Chapter 2, Arduino: The Brains of an Embedded System

> This chapter familiarizes you with Arduino, the open source electronics prototyping platform used in every project in this book (except the Boxing Clock in Chapter 6).

Chapter 3, Stalker Guard

> In this chapter, you'll learn how to use distance-finding sensors to detect when someone is trying to sneak up on you.

Chapter 4, Insect Robot

> This chapter uses distance-finding sensors, servos, and spare parts to make an obstacle-avoiding robot.

Chapter 5, Interactive Painting

> This chapter combines Arduino, your computer, and distance-finding sensors to create an interactive slideshow you can control with your hands. You'll also learn about two languages for programming on the computer: Processing and Python.

Chapter 6, Boxing Clock

> This chapter teaches you how to build a graphically rich timer clock on an Android phone. It will also serve as a primer for Chapter 8.

Chapter 7, Remote for a Smart Home

> In this chapter, you'll hack some remote-controlled power outlets so you can turn things on or off using a sketch running on Arduino—or even from the convenience of your desktop computer.

Chapter 8, Soccer Robot

This chapter combines a lot of what you've learned so far: Arduino, robotics, and cell phone (Android) programming. You'll learn how to create a remote-controlled, soccer-playing robot. You'll control it from your cell phone's built-in accelerometer; simply tilt the phone to tell the robot to move or kick a small ball!

Appendix, tBlue Library for Android

The appendix presents tBlue, a lightweight library that makes it easy to communicate over Bluetooth between an Android phone and Arduino.

Conventions Used in This Book

The following typographical conventions are used in this book:

Italic

Indicates new terms, URLs, email addresses, filenames, and file extensions.

Constant width

Used for program listings, as well as within paragraphs to refer to program elements such as variable or function names, databases, data types, environment variables, statements, and keywords.

Constant width bold

Shows commands or other text that should be typed literally by the user.

Constant width italic

Shows text that should be replaced with user-supplied values or by values determined by context.

Using Code Examples

This book is here to help you get your job done. In general, you may use the code in this book in your programs and documentation. You do not need to contact us for permission unless you're reproducing a significant portion of the code.

For example, writing a program that uses several chunks of code from this book does not require permission. Selling or distributing a CD-ROM of examples from O'Reilly books does require permission. Answering a question by citing this book and quoting example code does not require permission. Incorporating a significant amount of example code from this book into your product's documentation does require permission.

We appreciate attribution. An attribution usually includes the title, authors, publisher, copyright holder, and ISBN. For example: "*Make: Arduino Bots and Gadgets*, by Kimmo Karvinen and Tero Karvinen (O'Reilly). Copyright 2011 O'Reilly Media, 978-1-449-38971-0." If you feel that your use of code examples falls outside fair use or the permission given above, feel free to contact us at *permissions@oreilly.com*.

We'd Like to Hear from You

Please address comments and questions concerning this book to the publisher:

O'Reilly Media, Inc.
1005 Gravenstein Highway North
Sebastopol, CA 95472
(800) 998-9938 (in the United States or Canada)
(707) 829-0515 (international or local)
(707) 829-0104 (fax)

We have a website for this book, where we list errata, examples, and any additional information. You can access this page at: *http://oreilly.com /catalog/9781449389710*. All code examples and programs are available on *http://BotBook.com*.

To comment or ask technical questions about this book, send email to: *bookquestions@oreilly.com*.

Maker Media is a division of O'Reilly Media devoted entirely to the growing community of resourceful people who believe that if you can imagine it, you can make it. Consisting of *MAKE Magazine*, *CRAFT Magazine*, Maker Faire, and the Hacks series of books, Maker Media encourages the Do-It-Yourself mentality by providing creative inspiration and instruction.

For more information about Maker Media, visit us online:

MAKE: *www.makezine.com*
CRAFT: *www.craftzine.com*
Maker Faire: *www.makerfaire.com*
Hacks: *www.hackszine.com*

Safari® Books Online

 Safari Books Online is an on-demand digital library that lets you easily search over 7,500 technology and creative reference books and videos to find the answers you need quickly.

With a subscription, you can read any page and watch any video from our library online. Read books on your cell phone and mobile devices. Access new titles before they are available for print, and get exclusive access to manuscripts in development and post feedback for the authors. Copy and paste code samples, organize your favorites, download chapters, bookmark key sections, create notes, print out pages, and benefit from tons of other time-saving features.

O'Reilly Media has uploaded this book to the Safari Books Online service. To have full digital access to this book and others on similar topics from O'Reilly and other publishers, sign up for free at *http://my.safaribooksonline.com*.

Acknowledgments

Thanks to:

- Juho Jouhtimäki
- Marjatta Karvinen
- Nina Korhonen
- Mikko Toivonen
- Marianna Väre
- Medialab, Aalto University School of Art and Design
- O'Reilly Media
- Readme.fi
- Tiko, Haaga-Helia University of Applied Sciences

Introduction 1

This chapter will get you started building and designing prototypes for embedded systems. You will learn basic principles that you'll follow in Chapters 3 and 4 as you build the Stalker Guard and Robot Insect. Prototypes in this book are just the beginning. Once you know the techniques, you'll be able to build prototypes for your own inventions.

Building Philosophy

When you break a programming problem down into smaller pieces, be sure to test and validate each piece as you go. If you don't do this, you could find yourself wildly off track by the time you've gotten through a few pieces.

Prototype

This book provides techniques for building *prototypes*, or test versions of a device. A prototype such as the one shown in Figure 1-1 provides a proof of concept—a concrete realization of a device's intended functions.

Try to finish a functional prototype as quickly as possible. Once you've documented a working prototype, you can build in improvements in later versions.

You can make a working end result by stripping out unnecessary functions and taking shortcuts. If it makes testing quicker, use rubber bands and duct tape when you have to. Don't try to optimize your code in the first version.

It's much easier to build an impressive version once the first prototype is finished. Usually, you'll find that many challenging problems you face in the prototype don't even need to be solved for the final version. In the same way, building a prototype can reveal new opportunities for development.

Figure 1-1. *Jari Suominen testing a prototype made of Legos*

1

Having a prototype can also help you secure funding for your project. Who would you believe more: someone who talks about a walking robot, or someone who has actually built one?

Start with Hello World

Starting a project with Hello World is usually a good idea, because it's the simplest possible program. Typically, Hello World will print a row of text to a computer screen or blink an LED. It is used for testing to make sure the development environment works.

If your next, more complicated iteration doesn't work, you can search for the cause of the problem within the added code. Hello World lets you know that the microcontroller, development environment, interpreter, and USB port all function correctly.

Build in Small Steps

Complex problems (see Figure 1-2, Figure 1-3, and Figure 1-4) are hard to solve, but you can usually make them easier by breaking them down into smaller pieces. You can then solve the problem one manageable piece at a time.

A student of ours once built a burglar alarm after studying embedded systems for a week. The alarm buzzed whenever an infrared sensor detected movement. Users could log into the system wirelessly by presenting an ID in the form of a keychain. Once the system approved the login, the user could then move freely in the space without triggering an alarm.

> *How does Arduino say "hello" to the world? By blinking an LED. You'll learn more in Chapter 2.*

Figure 1-2. *Juho Jouhtimäki and Elise Liikala building a motion-sensitive soft toy*

Figure 1-3. *Welding a robot hand*

A project like this can sound quite complex to a novice, but it really consists of three clearly separate components (motion detector, buzzer, RFID reader). First, the student programmed and tested the motion sensor. That section was finished when the program could detect movement and sound the alarm.

The three components of the system do not affect one another in any way, and the only unifying factor is the code. Program code can check with the motion detector to determine whether movement is present and, if so, it can switch on the buzzer.

Figure 1-4. *Jari Suominen's strobo owl, which uses aesthetics from printed circuit boards and components*

Test in Steps

"I wrote the code for a singing and dancing robot that can walk up stairs. The code is 30,000 lines long. I just tried compiling it, but it doesn't work. Do you have any advice?"

Conduct testing as early as possible. If, for example, you build a walking robot, the first thing to test is whether you can make the servo motor move. The next test can make the servo move back and forth.

After you have tested the functionality of a specific version of code, save it separately from the version you are working on.

Revert to the Last Known Good Version

When you have developed your code into a confusing and nonfunctional state, the solution is easy. Go back to the last working version.

More specifically, go back to a working stage when the situation was already becoming confusing. This method removes the problem areas and lets you start over with a functional clean slate, helping you isolate what went wrong.

Read the Friendly Manual

RTFM is an old Internet acronym. (Actually, the *F* is not always *friendly*, so we usually stick with just RTM.) The point of the expression is that most answers are out there, written in a manual. When you're surrounded by parts (see Figure 1-5), you're going to need answers.

Friends and students sometimes wonder how we know so much. How do we know the Arduino operating voltage or the way to install SSL encryption to the Apache web server?

Figure 1-5. *Mikko Toivonen, surrounded by robots and microcontrollers*

The answer is easy. You can find instructions for almost anything if you know where to look.

Instructions don't always come with devices and parts, but you can often find them on manufacturer's web pages (such as *http://www.parallax.com*) or by searching in Google. Good search terms include device names (e.g., "ping ultrasonic sensor") or a sequence of numbers on a circuit board (e.g., "H48C").

You could also combine a search sequence with a technology—for example, "H48C arduino." Some web pages are devoted specifically to Arduino—for example, *http://arduino.cc* and our site, *http://BotBook.com*.

Document

Most things appear easy once you know them. The details of a project seem obvious on the day you complete them ("*of course* I remember when I programmed the 16-servo walker"). But a week after building, coding details begin to disappear from your memory. After a year or so, it can be hard for someone who builds many projects to remember anything about a specific one.

For this reason, it is worthwhile to document all projects. Typing notes avoids the potential problem of illegible handwriting, and shooting stages with a digital camera provides an accurate visual snapshot of each stage.

You might also consider publishing your results on the Web. Some projects that would otherwise be collecting dust in your drawer might actually be useful to others. You might even find your own instructions (long since forgotten) when looking to solve a new problem with similar logic. Two sites where you can publish projects are Make: Projects (*http://www.makeprojects.com*) and Instructables (*http://www.instructables.com*).

Figure 1-6. *Jenna Sutela and David Szauder demoing functions of a wearable prototype*

Reusing Parts

Prototype mechanics (see Figure 1-6) need all kinds of parts, such as frames, limbs, and joints. Finding appropriate materials can seem daunting. Customizing more complicated parts using homebrew methods isn't always easy, and even basic materials—such as lightweight and sturdy metal plates—can be significantly expensive at hardware stores.

Not every device is safe to salvage: for example, a CRT (Cathode Ray Tube) TV retains a hazardous voltage for a long time after you unplug it from the wall.

As a starting point, we recommend using recycled parts. Old devices are filled with usable materials, so remove all salvageable parts before you throw them away.

One additional perk that comes with using recycled parts is a unique aesthetic. Old parts often have interesting shapes, curves, and worn areas (Figure 1-7).

Figure 1-7. *An assortment of parts that can be reused*

Computer DVD drives and hard drives can make great frames for robots, because their covers are often made of lightweight, easily drillable, and sturdy material. You can also remove DC (direct current) motors and gears from DVD drives. Nowadays, there is more readily available computer junk than you can gather and store in your home. Educational institutions and corporations are particularly good sources, as they're continuously throwing out old devices.

Flea markets can also hold great finds. Mechanical typewriters deserve a special mention here. Though they are relatively hard to disassemble, they house an unbelievable amount of small springs, metal pieces of different shapes, and screws.

Disassemble devices as soon as you find them and then discard or recycle unnecessary parts. This way, you'll avoid turning your home into a graveyard of retired devices, and more importantly, the parts will be immediately usable when you really need them. When you are searching for a suitable attachment piece for a servo, you probably don't want to start a six-hour disassembly operation. Parts usually won't find a new purpose until you've removed them from the original device, at which point inspiration might strike. You might even wonder how a specific "whatchamacallit" fits a new purpose so perfectly.

When you begin working on some difficult new mechanism, think about where you might have seen something similar. You'll often find everyday solutions to many problems. For example, parts purchased from bicycle or automotive shops can sometimes work in other projects. Figure 1-8 shows a hand with fingers that are moved with servo motors; every joint in each finger bends. The fingers were made by attaching sections of a steel pipe to a bicycle chain. They bend when a brake cable is pulled down. Typewriter parts welded to the opposite side of the structure pull the fingers back into a straight position.

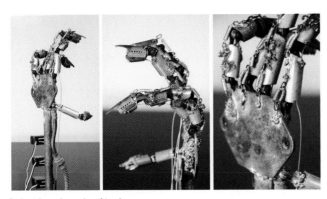

Figure 1-8. *Robot hand made of junk*

Also keep your eyes open in military surplus stores, where you can find inexpensive, sturdy, and personalized enclosures for prototypes. Various parts and accessories in these shops can also, with a bit of creativity on your part, give devices significantly more street cred. For example, Figure 1-9 shows a porcupine robot cover built from an MG/42 machine gun ammunition belt.

Figure 1-9. *Porcupine robot cover made from a machine gun ammunition belt*

Buying Components

If you can't find exactly what you're looking for in recycled materials, order component parts online. Many unique components can't be found locally at all, or will be overpriced if you do find them. Luckily, comparing prices and ordering online is quite easy.

Because online stores often change, make sure to check the latest links available at *http://BotBook.com/*.

We purchased parts for this book from a variety of sources. We ordered most of the sensors and full-rotation servos from the United States. Arduinos and some of the sensors came from Sweden. We rounded up ordinary components—such as resistors, LEDs, and wiring—from electronics stores in Helsinki. Standard servos came from a Finnish online store specializing in radio-controlled cars and airplanes. Some servos were ordered from Hong Kong. Here are a few sources to consider:

Maker SHED

    MAKE Magazine's store can be found online at *http://www.makershed .com/* and in real life at Maker Faire (*http://makerfaire.com/*). Maker SHED carries Arduinos, project kits, tools, parts bundles, books, and much more. Keep on eye on Maker SHED for special parts bundles or kits dedicated to projects in this book.

Adafruit Industries

    The Adafruit store (*http://www.adafruit.com*) specializes in Arduinos, microcontrollers, electronic and robotic components (including servo motors), tools, and kits. It also has a comprehensive set of Arduino tutorials and produces its own Arduino-compatible boards such as the Boarduino.

SparkFun Electronics

    Among many other things, SparkFun (*http://www.sparkfun.com/*) is a great source for all kinds of sensors—from light and temperature sensors to accelerometers and gas sensors. What's more, it sells the sensors mounted to breakout boards so you can easily connect them to an Arduino without having to do tricky surface-mount soldering. SparkFun has much more, including tools, parts, and Arduinos.

If you're in the US, you will generally be able to find all the parts you need within the country. However, if you ever need to order large amounts of something (such as hundreds or thousands of LEDs), you may find yourself purchasing from an overseas supplier (for example, many bulk LED sellers on eBay ship from Hong Kong).

When choosing a country to order from, take into consideration customs rules and additional fees incurred by international orders. Shipping costs can also be high in some countries, and some companies won't even ship overseas. Also, consumer protections might not apply to international orders in the event that the package is broken or the product is different from what you ordered.

Regardless of all the scaremongering, ordering internationally usually works out without major problems. We have received everything we have ordered, and the products haven't had any major faults.

Chapter 1

Useful Tools

When building prototypes, you're going to need some tools (Figure 1-10). The following sections cover the tools that we have found a consistent need for. They are not all mandatory, but depending on your own projects or needs, you may have a use for them in the future.

Hearing Protectors and Safety Glasses

When using power tools, you must cover your ears with proper hearing protectors and wear safety glasses to protect your eyes from harmful flying debris and material fragments (Figure 1-11). Note that metal can fly forcefully, even when you're cutting or bending with pliers.

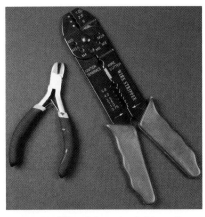

Figure 1-10. *Wire stripper and side-cutter pliers are sufficient for building prototypes on a prototyping board*

Figure 1-11. *Hearing protectors and safety glasses*

Needlenose Electronics Pliers

You should immediately purchase good needlenose pliers (Figure 1-12), which can be used to grab small components and parts. The tip for the pliers should be sharp enough to fit into even the smallest of spaces.

Figure 1-12. *Needlenose electronics pliers*

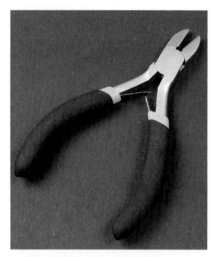

Figure 1-13. *Diagonal-cutter pliers*

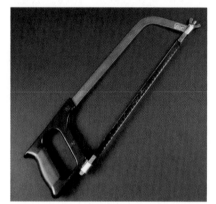

Figure 1-14. *Metal saw*

Diagonal-Cutter Pliers

Diagonal-cutter (or side-cutter) pliers, shown in Figure 1-13, are used for cutting wires and are also suitable for other small cutting jobs. Always keep at least one set of side cutters in good shape, and use a secondary pair for tasks that cause more wear.

Metal Saw

A metal saw is a basic, functional tool for shaping and cutting metal (Figure 1-14). Keep a spare blade on hand to keep promising building processes from being interrupted by a broken blade.

Wire Strippers

Wire strippers are used to remove the plastic around a wire to expose a conducting metal within specific areas. Do not use your teeth to strip wires! It is much more expensive to fix dental enamel than to spend just a few dollars on good wire strippers. The adjustable wire strippers on the left side of Figure 1-15 are much more useful than the multigauge model on the right, but they're not as common.

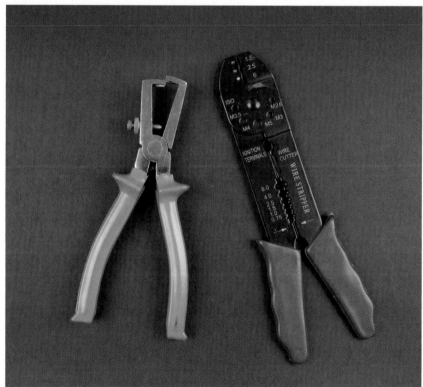

Figure 1-15. *Wire strippers*

Screwdrivers

You'll need many different types of screwdrivers, especially when opening devices. Using the wrong screwdriver tip for a particular screw could destroy either the screw or the screwdriver and is just not worth the potential damage. The easiest and most economical thing to do is to buy a kit that comes with a handle and various attachable bits (Figure 1-16). Many electronic devices require a Torx driver and can't be opened with a flat- or Phillips-head screwdriver.

Alligator Clips

Alligator clips (Figure 1-17) can be useful for quickly connecting components and cables. They can also connect multimeter probes, enabling hands-free measurements.

Electric Drill

You'll need an electric drill for many projects. A hammer drill, shown in Figure 1-18, is also suitable for drilling into concrete, but a rechargeable cordless drill is easier to handle.

A drill bit can break easily, especially when you're drilling metal with thin bits, so you must wear eye protection when working with a drill. Always position the drill directly into the hole; drilling at an angle will bend the bit and cause it to break under rotation.

Figure 1-16. *Screwdriver kit with a variety of bits*

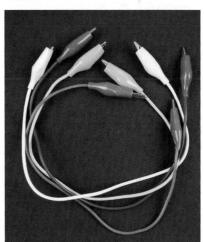

Figure 1-17. *Alligator clips*

Figure 1-18. *Electric drill*

Figure 1-19. *Leatherman*

Leatherman

A portable handy tool such as a Leatherman (Figure 1-19) is useful during several phases of project building. In this case, it makes sense to invest in the name-brand tool rather than buying cheap imitations. A high-quality multipurpose tool can withstand heavy use, and its individual parts function in the same way as separate tools.

Maker SHED sells an assortment of MAKE-branded Leatherman Squirt tools, such as the MAKE: Circuit Breaker Leatherman, a set of electronics tools that can fit on a keychain. See http://www.makershed.com/SearchResults. asp?Search=leatherman *for more information.*

Mini Drill

A mini drill (Figure 1-20) is not absolutely necessary, but it makes many tasks easier. Compared to an electric drill, a mini drill is lightweight and relatively precise to work with.

By using an appropriate bit, you can use a mini drill for drilling, sanding, sharpening, shining, cutting, and more. Of course, it doesn't replace a normal drill, because it doesn't have sufficient torque for drilling larger holes.

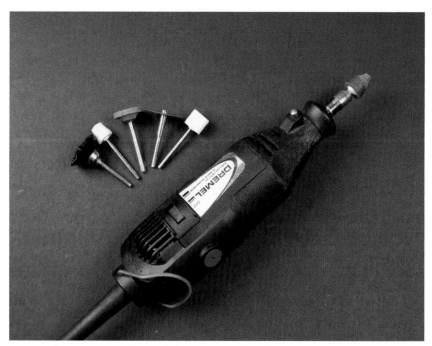

Figure 1-20. *Mini drill*

Headlamp

A headlamp (Figure 1-21) can be handy for focusing light in the direction you're working. Additional light is useful to have, even in well-lit spaces.

Hot-Glue Gun

A hot-glue gun (Figure 1-22) can adhere items together quickly. The resulting connection is not necessarily very strong, and glued items can bend away from each other, but it works sufficiently well in many prototyping phases. In addition, the fact that hot glue hardens quickly, and items glued with it can be (at least in theory) removed from each other relatively easily, can make the building process less stressful. Still, hot glue is not a replacement for Blu-Tack, and another downside is that if you're unsuccessful in your first attempt to join items together using hot glue, you'll usually need to scrape and shine the surfaces before trying again.

Figure 1-21. *Headlamp*

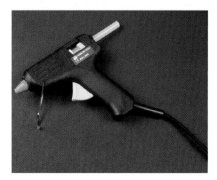

Figure 1-22. *Hot-glue gun*

Nail Punch and Hammer

Drilling metal at home without a drill press can be quite challenging, especially with smooth metal surfaces on which a bit can slide and go through the wrong spot. A nail punch (Figure 1-23, left) can fix this problem. It can create a small dent on the spot where you want to drill a hole, making drilling much easier.

A hammer is a useful tool in its own right, but it's not always the right tool for the job. If you have something to dislodge or to set in place, look for a gentler tool first, so you don't break your project into many little pieces. As Abraham Maslow said, "I suppose it is tempting, if the only tool you have is a hammer, to treat everything as if it were a nail."

Figure 1-23. *Nail punch and hammer*

Soldering Iron

A soldering iron (Figure 1-24) joins metal sections of components together with molten metal (usually lead, but lead-free solder is available as well). The tip of a soldering iron must be sufficiently thin to enable precise attachment of small parts. Irons with a built-in thermostat are more expensive, but having the capability to adjust the temperature lessens the likelihood of destroying more sensitive components. You will learn the basics of soldering in Chapter 3.

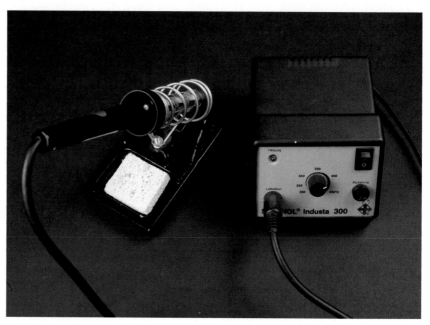

Figure 1-24. *Soldering iron*

Multimeter

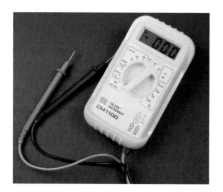

Figure 1-25. *Multimeter*

A multimeter (Figure 1-25) is used for measuring current, voltage, and resistance. You can use it to test a value of a resistor or whether two sections of a circuit are connected. You also can test the condition of a battery by measuring its voltage.

The multimeter shown in Figure 1-25 has two ranges for measuring voltage: DC (direct current) and AC (alternating current). All Arduino circuits in this book use direct current. The correct measurement range for voltage and resistance is the smallest possible range onto which measured readings can fit.

A continuity test works technically in the same way as measuring a value of a resistor. Instead of displaying a resistance value, the continuity test beeps when an unrestricted flow of electricity is detected between two measurement probes.

Figures 1-26 and 1-27 illustrate some common uses for a multimeter. The Interactive Painting project in Chapter 5 covers measuring resistance in more detail.

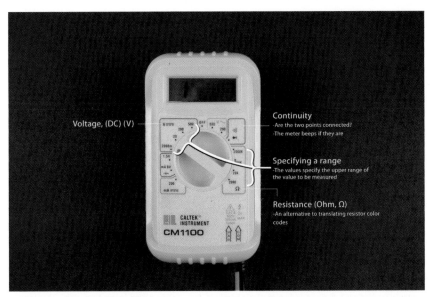

Figure 1-26. *The most common functions of a multimeter*

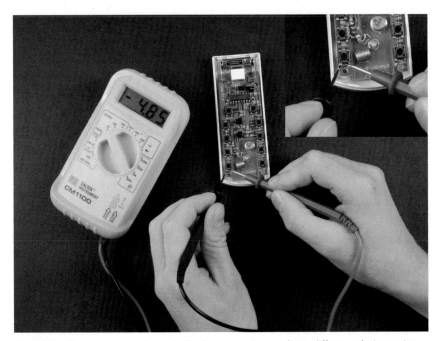

Figure 1-27. *Studying a remote controller by measuring a voltage difference between two terminals of a button*

Electronic Circuit Theory Review

We'll end this chapter with just enough theory to get you started with the practice.

Voltage Creates an Electrical Current

Voltage refers to a difference in electrical potential between two parts of a circuit. For example, the terminals of a battery can have a 9-volt voltage between them.

If two parts of a circuit with different electrical charges are connected, voltage potential creates a current flow. For example, current will start flowing through a lamp that is connected between the two terminals of a battery, causing the lamp to light up.

A unit of voltage is a *volt* (V). The Arduino microcontroller used in this book functions with a minimum 7V and maximum 12V power adapter (or it can be powered from a 5V USB connection). Voltages inside computers are within a similar range. US AC sockets provide 110 volts and European AC sockets provide 230 volts.

A lamp will be brighter with a 9V battery than with a smaller 4.5V battery. Larger voltage creates a larger current. If a component is used with a voltage higher than what it is rated for, it will usually burn out. If you supply 5 volts to an LED that is rated for 2.4 volts, it will probably make a popping sound, release a little smoke, and cease to function. A running joke among electrical engineers and technicians is that once you've released the "magic smoke" inside an electronic component, you can't put it back in.*

A Resistor Resists the Flow of Current

If a resistor is added between a lamp and a battery, the lamp will be dimmer. A resistor resists the flow of current.

All components create at least a bit of resistance. A filament of an incandescent light bulb is sufficient by itself to resist the current flow.

A resistor may be all that's needed to avoid releasing the magic smoke inside an LED. For example, a 1 kOhm resistor is generally more than sufficient to protect a red LED. If you have the specifications for your LED, you can calculate the value of the resistor.

Evil Mad Scientist Laboratories has a handy papercraft pocket LED calculator that you can print out and carry with you: http://www.evilmadscientist.com/article.php/ledcalc.

---

* *http://en.wikipedia.org/wiki/Magic_smoke*

Chapter 1

Short Circuits Are Dangerous

If you bridge a battery's positive and negative terminals with a wire, it forms a *short circuit*. The current flows rapidly through the wire, and the wire and battery will both become warm and may possibly leak or explode. Why do we mention this? Because it's possible to create a short circuit in your own projects if you don't use the correct resistor values. When you follow the instructions to build a project, you must be sure to use the resistor values specified to avoid creating the hazardous condition that comes with a short circuit.

Closed Circuits Allow Electricity to Flow

When a device is powered, its circuit is closed and electricity will flow through the device. An *open circuit* means that electricity cannot flow through a device. For example, a device that is shut down by its power switch is an open circuit. Electricity can't flow when the circuit is opened by the switch.

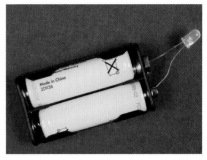

Figure 1-28 shows a closed circuit: two batteries powering an LED. The magic smoke didn't come out because the batteries and LED are well matched: the LED has a voltage of 2.6V, which is more than the voltage delivered by two AA rechargeable batteries. Standard AA batteries (1.5V each) might overpower the LED. Still, if you intended to run this circuit for hours on end, it would be advisable to include a low-rated resistor, even a 10 or 100 Ohm.

Figure 1-28. Simple closed circuit powering an LED

Ground = Zero Voltage Level

To make it easier to discuss topics related to voltage, a single point in a circuit is usually compared to the negative terminal of a power supply. The voltage level of a negative terminal is 0V, against which all other points of the circuit are measured. For example, the positive terminal of a 9V battery can be said to have 9V of voltage.

Figure 1-29. Symbol for ground

Ground has many names, all of which mean the same thing: *0V*, *minus terminal*, *earth*, and *GND*. Black wire is often used to connect to the ground (red is used for positive voltage). In a circuit, ground is marked with its own symbol (shown in Figure 1-29) to avoid having to always draw a line to the minus terminal.

In this chapter, we've covered prototyping principles, techniques, and tools, and reviewed some basics of electrical theory. Now we'll move on to Chapter 2, where we introduce Arduino, the open source prototyping platform that will be the brain of your projects.

Arduino: The Brains of an Embedded System

2

In this chapter, you'll compile a program you have written onto an Arduino microcontroller, a small computer that acts as the brains of an embedded system. Arduino, an easy-to-learn hardware and software development environment and prototyping platform, is the foundation for the projects we'll complete in upcoming chapters.

A *microcontroller* is a small computer with a processor and memory that controls the functions of many everyday devices. Some microcontrollers are designed to connect easily to a computer for programming for specialized purposes. Arduino is an example of one of these easy-to-program microcontrollers.

Microcontrollers make it easier to build electronic devices because you can control their functions via code. Microcontrollers can control and interpret forms of both input and output. For example, you can flicker an LED by connecting it to a specific Arduino pin with code that instructs it to switch the current on for one second and then off for one second. The LED is an example of an output, which you could then control using a sensor, button, switch, or any other form of input. Naturally, most programs do many other, more sophisticated tasks. Microcontrollers enable us to solve quite complex problems step by step.

Why Arduino?

The most suitable microcontroller choices for a beginner are Basic Stamp and Arduino. Basic Stamp has existed since the early 1990s and has become popular among hobbyists. It uses the Basic programming language, which is easy to use but somewhat limited compared to the C language used by Arduino.

Functionally, Arduino is quite similar to Stamp, but it solves many problems that Stamp has traditionally faced. One significant feature for hobbyists is Arduino's lower cost: the basic Arduino starting package is approximately a quarter of the price of a comparable Stamp package. And, despite its cheaper price, Arduino has a more powerful processor and more memory.

Arduino is also smaller than Stamp, which is beneficial in many projects. The Arduino Uno model (see Figure 2-1, left) is slightly smaller than the Stamp, but the tiny Arduino Nano (Figure 2-1, right) is about the same size as the Stamp *module* that sits on the Stamp board (just above the serial port in Figure 2-2). For comparison, Figure 2-2 shows the Stamp and the Nano next to each other.

Figure 2-1. *Arduino Uno (left) and Arduino Nano (right)*

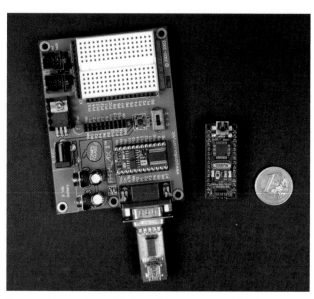

Figure 2-2. *Basic Stamp (left) and Arduino Nano (right)*

One final asset is that the Arduino programming environment is based on open source code and can be installed on Windows, Mac OS X, and Linux.

Starting with Arduino

Chapter 8 includes a project that uses Bluetooth. Although there is an Arduino model with built-in Bluetooth (Arduino BT), a more flexible option when you're creating Bluetooth projects with Arduino is to use a third-party Bluetooth adapter, such as SparkFun's Bluetooth Mate (http://www.sparkfun.com/products/10393). This will allow you to use the Bluetooth module with different projects, or to replace Bluetooth in one of your projects with another type of wireless module such as an XBee radio.

Arduino is available in a few different models. This book covers the aforementioned Arduino Uno and Arduino Nano. Uno is an inexpensive (around $30) and sturdy basic model, and is the most current version of the board. It was released publicly in September 2010 and is the successor to the Arduino Diecimila and Arduino Duemilanove. Nano is significantly smaller, but more fragile and slightly more expensive ($35). Both models are described in a bit more depth at the end of this chapter.

First, you have to buy an Arduino and a compatible USB cable. Uno and Nano communicate to your computer via USB (for uploading new programs or sending messages back and forth). They can also take their power over USB. Uno uses a USB-B cable and Nano uses a Mini-B, and each connects to the computer with a USB-A male connector. All three connectors are shown in Figure 2-3.

Figure 2-3. *Arduino USB cables: Mini-B, USB-A, and USB-B*

Installing Arduino Software

Next, you need to install the Arduino development environment for your operating system and compile the first test program. This "Hello World" code is the most important part of getting started with a new device. Once you are able to compile simple, light-blinking code in Arduino, the rest is easy.

The examples in this book were tested with version 0021 of the Arduino development environment. If you decide to use some other version, the installation routine might differ. If you are using an operating system other than Windows, Ubuntu Linux, or Mac OS X, or an Arduino other than Uno or Nano, look for installation instructions at *http://arduino.cc/*. And remember that you will find all complete code examples, links, and program installation packages at *http://BotBook.com/*.

Windows 7

Here's how to get up and running under Windows 7:

1. Download the Arduino development environment from *http://arduino.cc/en/Main/Software* and unzip it to the desired folder by clicking the right button and selecting "Extract all."

2. Connect the USB cable to your computer and to the Arduino's USB port. The Arduino LED should light green.

3. Windows will search for and install the necessary drivers automatically. It notifies you when the installation is complete. If Windows does not locate the driver:

    a. Open Device Manager by clicking the Start Menu, right-clicking Computer, choosing Properties, and then clicking Device Manager in the list of options on the left.

b. Locate Arduino Uno in the list of devices (it should be in the section called Other Devices). Right-click it and choose Update Driver Software.

c. Choose "Browse my computer for driver software."

d. Navigate to the Arduino folder you extracted, select the *drivers* subdirectory, and press Next.

e. If prompted to permit the installation of this driver, choose "Install this driver software anyway."

When the driver is successfully installed, you'll see the dialog shown in Figure 2-4.

Figure 2-4. *Drivers installed*

Windows XP

In general, installation for most Windows XP programs is pretty similar to Windows 7, but Arduino is an exception. If you have XP, start by downloading the Arduino development environment, extracting the file to a location on your computer, and connecting the Arduino to your computer as described in the previous section. Then follow these additional instructions:

1. Windows opens the Found New Hardware Wizard.

2. Select "Install from a list or specific location" in the window and press Next.

3. Deselect the checkbox in "Search removable media" and check the box "Include this location in the search." Navigate to the Arduino folder you extracted, select the *drivers* subdirectory, and press Next. If you are using an older model of Arduino, or the Nano, you may need to choose the *drivers/FTDI USB Drivers* subdirectory instead.

4. Click Finish.

Ubuntu Linux

Though you can install Arduino on Ubuntu and other Linux environments using graphical user interface tools, the following steps use the Terminal (Figure 2-5) to simplify the instructions.

Open the Terminal by choosing Applications→Accessories→Terminal. The dollar sign at the beginning of the following command lines is the command prompt created by the computer; do not type the dollar sign, just the characters that follow it.

We tested this installation process with Ubuntu 9.04, but it should also function (with minor alterations) with other versions.

> *Before you try to install Arduino on Linux, consult the Arduino FAQ (http://arduino.cc/en/Main/FAQ#linux) for links to the latest instructions.*

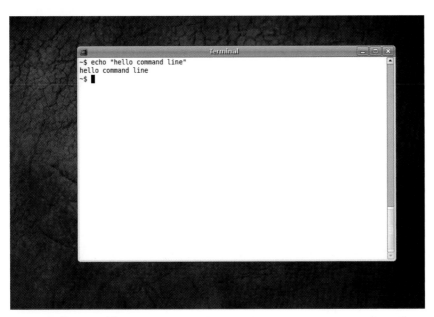

Figure 2-5. *Command line*

Start using the universe program repository, which includes free, open source programs, with publicly available source codes:

```
$ sudo software-properties-gtk --enable-component=universe
```

When asked by sudo, type your password. The command after sudo will be executed using root user privileges.

Update the available software list:

```
$ sudo apt-get update
```

Now it's time to install *dependencies*: all the programs the Arduino development environment requires to function, including Java (*openjdk*) and programming tools for the AVR chip *gcc-avr*, *avr-libc*, and *avrdude*. New 64-bit computers also require the 32-bit compatibility library *ia32-libs*.

```
$ sudo apt-get install --yes gcc-avr avr-libc avrdude openjdk-6-jre
$ sudo apt-get install --yes ia32-libs
```

Next, download and open the Arduino development environment from the official Arduino home page (*http://arduino.cc/en/Main/Software*), where you'll find two packages: "Linux (32bit)" and "Linux (AMD 64bit)." Newer computers are based on 64-bit technology. If you don't know which package to download, use the `uname-m` command to determine whether your computer is a newer 64-bit model (*x86_64*) or an older 32-bit model (*i386*).

Uncompress the software package you downloaded (this will create an *arduino-version* directory under your current working directory):

```
$ tar -xf ~/Downloads/arduino-*.tgz
```

Start the Arduino development environment. Because you will execute the command from a specific folder, define the whole path to that folder:

```
$ ./arduino
```

The Arduino development environment will start.

Mac OS X

Here's how to get up and running under Mac OS X:

1. Download the Arduino development environment from *http://arduino.cc/en/Main/Software* and open the *.dmg* file.

2. A new Finder window appears with three icons (Arduino, a link to your Applications folder, and the FTDI USB serial driver package).

3. Drag the Arduino icon to your Applications folder.

4. If you are using a version of Arduino prior to the Uno, install the *FTDIUSB-SerialDriver* package.

5. When you connect the Arduino, you may see the message "A new network interface has been detected." Click Network Preferences and then click Apply. You can close the Network Preferences when you are done.

Hello World with Arduino

Now you're ready to upload your first Arduino program. Open the Arduino development environment:

Windows

Double-click the Arduino icon (you'll find it inside the *Arduino* folder that you extracted earlier).

Linux

Change directory to the Arduino folder and run Arduino:

```
$ cd arduino-0021
$ ./arduino
```

Mac OS X

Double-click the Arduino icon (you'll find it inside the *Arduino* folder).

Select Tools→Board→Arduino Uno, as shown in Figure 2-6. If you are using a different model of Arduino, select it instead.

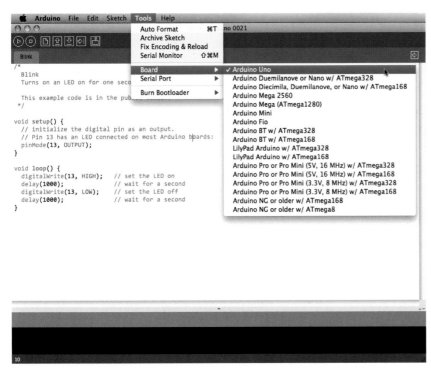

Figure 2-6. *Arduino board selection*

Select File→Sketchbook→Examples→1. Basics→Blink. This example code flashes the LED on the Arduino pin 13 (the Uno includes an onboard LED connected to pin 13).

Determine which serial port Arduino is using:

Windows

Open the Start menu, right-click the computer icon, and select Properties. System Properties will open. On Windows XP, click Hardware. On Vista or Windows 7, look in the list of links to the left. Select Device Manager from the list and open the "Ports (COM & LPT)" node. See which COM port is marked as a USB COM port, as shown in Figure 2-7.

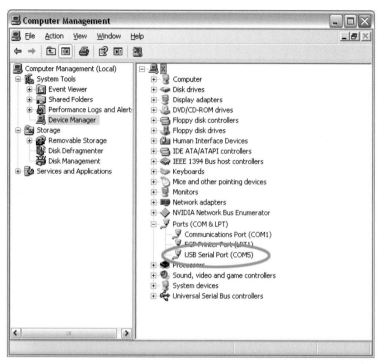

Figure 2-7. *Determine the correct port*

Linux and Mac OS X

With the Arduino unplugged from your computer, select Tools→Serial Port in the Arduino development environment, and observe the list of serial ports listed (such as */dev/ttyUSB0* in Linux or */dev/tty.Bluetooth-Modem* in Mac OS X). Dismiss the menu by clicking elsewhere onscreen.

Plug the Arduino in and choose the same menu options again (Tools→Serial Port). Observe which serial port appeared. You've figured out which serial port your Arduino is using.

In the Arduino development environment, select Tools→Serial Port and choose the port you found in the previous step.

Click the icon with the right-pointing arrow in a square box, or choose File→"Upload to I/O Board." The Arduino transmission lights will flash briefly, and you should see the message "Done uploading."

Now you should see the yellow LED labeled L on the Arduino board flashing (see Figure 2-8). This means you have successfully installed the Arduino development environment and uploaded your first Arduino program to the microcontroller. If the light is not flashing, follow the instructions again to see where the installation went wrong. You cannot proceed if this does not work. If you continue to have problems, see the online Arduino troubleshooting guide at *http://www.arduino.cc/en/Guide/Troubleshooting*.

Figure 2-8. *The LED on pin 13*

You might want to come back to this section and run through these steps if you have problems working with Arduino in the future; the Blink example is a good test to ensure that Arduino is working. If the LED does not flash, it's a good time to make sure the cord is plugged in, or determine whether your computer or the Arduino is having problems. Once you establish that something as simple as blinking an LED works, it will be easier to solve the more complex problems later.

Structure of "Hello World"

You just took an important step by running the "Hello World" of Arduino: Blink. The program blinks an internal LED on the Arduino. If you can get the LED to blink, you can be confident that you can compile and upload programs to Arduino.

All Arduino programs have a similar structure. Since Blink is a simple program, it is easy to understand the program structure by examining it.

Here is the source code for Blink, which comes with Arduino but is offered here with our commentary:

```
/*
  Blink ❶
  Turns on an LED on for one second, then off for one second, repeatedly.

*/

void setup() { ❷
  // initialize the digital pin as an output.
  // Pin 13 has an LED connected on most Arduino boards:
  pinMode ❸ (13, OUTPUT ❹); ❺
}
```

```
void loop() {  ❻
  digitalWrite(13, HIGH);  ❼     // set the LED on
  delay(1000);  ❽                // wait for a second
  digitalWrite(13, LOW);  ❾      // set the LED off
  delay(1000);                   // wait for a second
}
```

Let's review each section of the code.

❶ A slash followed by an asterisk (/*) opens a block of comments (they are ended by a */). Two forward slashes (//) indicate that the rest of the line is a comment. Comments consist only of information for the user; Arduino does not react to them. Comments will be removed during the compiling process, and they are not uploaded into the Arduino. This comment states the title and purpose of the program (known as a *sketch*). Most comments describe the purpose of a specific line or block of code.

❷ The setup() function executes once in the beginning of a program, where one-time declarations are made. The setup() function will be called automatically immediately after the Arduino is powered or has been programmed. Calling a function consists of a return value of a function (void), the name of the function (setup), and a list of parameters in parentheses. This function does not take parameters, so there is nothing inside the parentheses. This function does not return values, so its type is void (empty). When calling a function, commands are listed within a block of code, which is enclosed in curly braces ({ }).

❸ This setup() function includes only one command, which sets the pin connected to the LED into *output mode*. It calls the pinMode() function, which is defined in the Arduino libraries.

❹ OUTPUT is a constant defined within Arduino. When using digital pins on an Arduino, you must always declare them to be in either OUTPUT or INPUT mode.

> *An easy way to familiarize yourself with new functions is to highlight the name of the function (always in orange text), right-click, and select "Find in Reference." Here you will find an explanation of usage, syntax, and an example.*

❺ As is required by the C language (which Arduino is based on), a line ends with a semicolon.

❻ The majority of the execution time of the program will repeat the loop() function. The loop() function is called automatically (and repeatedly) after the setup() function finishes.

❼ To address a digital pin that has been set to be an OUTPUT, we use the digitalWrite() function. Set digital pin 13 on HIGH (which means +5V). Digital pin 13 is unique in that Arduino has a built-in LED and resistor attached to it, specifically for debugging purposes. The LED will light up. The pin remains on HIGH until we address the pin with another call from digitalWrite().

❽ Wait 1,000 milliseconds (1,000 one-thousandths of a second, totaling one full second). The LED is lit all the time.

❾ Switch the pin off (LOW) and wait again for a second. This is the last command in the loop() function, which means that the execution of the loop() function ends.

The program will call the loop() function over and over again automatically. The execution will continue from the first line of the loop() function, which sets the LED (ledPin) on HIGH. The execution of the program continues by repeating the loop() function until it is stopped by disconnecting power from the Arduino.

Arduino Uno

The Arduino Uno (Figure 2-9) is a good choice for your first Arduino. It is inexpensive and reliable. The Uno will use the power provided by the USB cable when the cable is connected to a computer. If necessary, it can also be powered by an external power supply such as a battery. When you upload code from your computer, the program is saved to the microcontroller itself. This means you can disconnect the Arduino and allow it to function as an independent device.

The Uno's pins have female headers that enable you to connect wires without soldering. This speeds up the building of simple prototypes but is not a very good longer-term solution, because cables can fall off relatively easily.

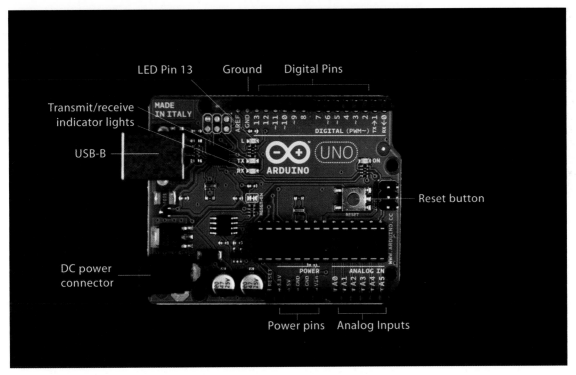

Figure 2-9. *Arduino Uno*

Arduino Nano

The Arduino Nano (Figure 2-10) is considerably smaller than the Uno mentioned earlier. It also has pins that you can connect straight onto a prototyping breadboard. These allow you to easily construct even quite complex circuits without soldering.

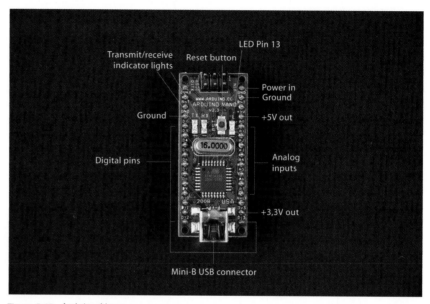

Figure 2-10. *Arduino Nano*

Nano is more expensive and sensitive than Uno. For example, a certain kind of short circuit will break Nano permanently. Another downside is that it is harder to read the markings on the pins, making it easier to misplace wires.

With the addition of a mini breadboard and retractable USB cable, the Arduino Nano becomes part of a handy travel pack (see Figure 2-11).

> *The Arduino Pro Mini (see Chapter 8 for more details) is even smaller than the Nano. It does not include a USB-serial adapter onboard, so you need to use a separate USB-serial adapter to program it. However, it is extremely small and lightweight, and comes in low-power variants, which makes it an ideal choice for projects where weight is a significant issue, such as airborne drones or remote-controlled vehicles.*

Now you know the basics to get started with your first project. In the next chapter, you will try Arduino in practice while building the Stalker Guard.

Figure 2-11. *Arduino travel pack*

Chapter 2

Stalker Guard | 3

In this project, you will build a Stalker Guard (Figure 3-1), a simple alarm device that measures the distances of objects behind you and vibrates when something comes too close (see Figure 3-2). You will also learn a program you can easily modify for other projects that works by monitoring data sent by a sensor and reacting when specific conditions are filled.

Over the course of this project, you will learn the basics of Arduino programming, distance measurement with ultrasonic sensors, and motor control. The Stalker Guard can be easily customized into new variations by changing the values of the ultrasonic sensor, replacing the sensor with another one, or replacing the motor with, say, a speaker. With little effort, you can develop the circuit further—for example, by turning it into a height meter.

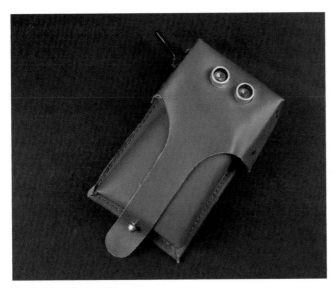

Figure 3-1. *The finished Stalker Guard*

Before starting the project, you'll need to install the Arduino development environment and make sure you can run the "Hello World" program from Chapter 2. The project is organized in steps, so you can always return to the previous step if you find that something isn't functioning correctly. Remember that you can download the complete code examples for each stage at *http://BotBook .com/* or *http://examples.oreilly.com/0636920010371*, which allows you to test their functions before reading the individual descriptions of each program.

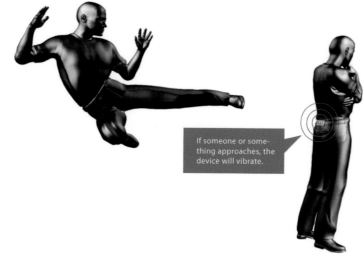

If someone or something approaches, the device will vibrate.

Figure 3-2. *The Stalker Guard in action*

What You'll Learn

In this chapter, you'll learn how to:

- Measure distance using the PING))) ultrasonic sensor
- Use a motor
- Apply the principles of Arduino programming
- Power Arduino from a battery

Tools and Parts

You'll need the following tools and parts for this project (Figure 3-3).

> *Manufacturer part numbers are shown for:*
> - *Maker SHED (US: http://makershed.com): SHED*
> - *Element14 (International and US; formerly Farnell and Newark, http://element-14 .com): EL14*
> - *SparkFun (US: http://sparkfun.com): SFE*

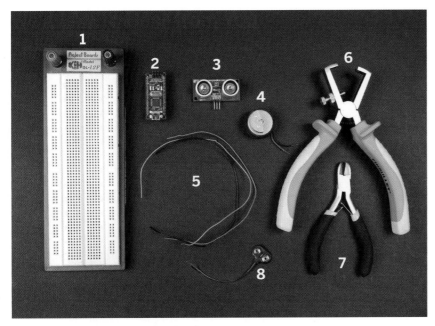

Figure 3-3. *Parts and tools used in this chapter*

1. Solderless breadboard (SHED: MKEL3; EL14: 15R8319; SFE: PRT-00112).

2. Arduino Nano (SHED: MKGR1; *http://store.gravitech.us*; or *http://store.gravitech.us/distributors.html*).

3. PING))) ultrasonic sensor (SHED: MKPX5; *http://www.parallax.com/Store/*).

4. Vibration motor (SFE: ROB-08449). If you can't find a vibration motor anywhere, you can replace it with an LED. These motors can also often be salvaged from broken cell phones.

5. Jumper wires, at least three colors (SHED: MKEL1; EL14: 10R0134; SFE: PRT-00124).

6. Wire strippers (EL14: 61M0803; SFE: TOL-08696).

7. Diagonal cutter pliers (EL14: 52F9064; SFE: TOL-08794).

8. 9V battery clip (EL14: 34M2183; SFE: PRT-00091).

Solderless Breadboard

To prototype circuits without soldering, you can use a solderless breadboard. To connect components, you'll push their legs (leads) into holes on the board. In Figure 3-4, two pairs of bus strips (power rails) are connected *horizontally* on the top and bottom of the board. These are typically used for providing power (positive voltage, often labeled VCC; and ground, labeled GND) access to the entire board. The terminal strips (center area of the board) are connected *vertically*. This area is where you will mount most components. You can remove parts from the board by gently pulling them out, which makes it easy to change circuits.

Note the power terminals at the left end of the breadboard in Figure 3-4. These are useful, but you won't find them in every breadboard. If your breadboard does not have power terminals, you can push power leads into any of the holes in the appropriate horizontal row to provide power to all the other leads along that row. We won't use power terminals in this project.

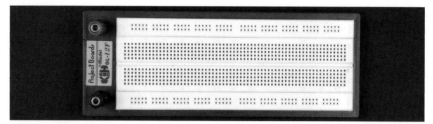

Figure 3-4. The holes in the prototyping breadboard are connected to each other by vertical rows; bus strips for power are connected horizontally

Figure 3-5 shows the breadboard with its bottom cover removed (don't do this to your breadboard, as it tends to damage it) so you can see more clearly how the rows are connected to each other.

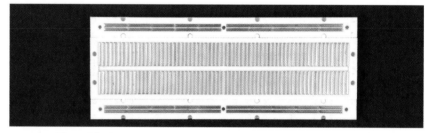

Figure 3-5. View of the bottom of the prototyping breadboard; metal plates conduct current from hole to hole

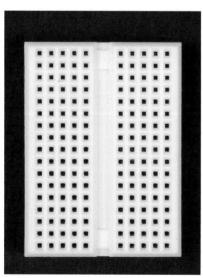

Figure 3-6. Small prototyping breadboard

So how would you connect the Arduino Nano's D5 (digital pin 5) pin to the second wire of a motor? Insert the Arduino into the middle of the prototyping board (also known as the "gutter"), so that it is straddling the vertical rows between both sides of the Arduino legs. This prevents us from short-circuiting the opposite side legs of the Arduino.

Push the Arduino into the board so that the Arduino's D5 pin is connected to each hole of the strip on the same side of the gutter (highlighted in Figure 3-9, which appears later in this chapter). Then simply insert the motor wire in the hole next to the Arduino D5 pin. Now the motor wire and Arduino pin D5 are connected to each other.

It's always a good idea to begin a project by building circuits on a prototyping breadboard. Once you're sure that a project functions correctly, you can then consider a more permanent place for it. Even in the next stage, you don't need to solder things together. For example, you can leave the more expensive parts, such as the Arduino itself, on a smaller prototyping breadboard (Figure 3-6). This way, you can easily use it in other projects and then put it back in its original place.

Jumper Wire

We're going to use jumper wire—a thin, single-strand electrical wire—to make connections to the prototyping breadboard. You'll need an adequate amount of visible conductive metal to connect the jumper wire deeply enough; about ¼ inch of exposed wire is sufficient.

Jumper wires (also known as *hookup wire*) are sold in ready-cut assortments (Figure 3-7), but you can also cut and strip them yourself from suitable wire. Generally, 22AWG solid-core wire is best. Using a different-colored wire for each distinct purpose can be useful. For example, use a red wire for power, and black for ground (0V, GND). Wires hooked to data pins often use other colors, such as blue or green. This method makes the circuit easier to comprehend.

Ping Ultrasonic Sensor

An ultrasonic sensor functions on the same principles as radar: it transmits a high-frequency signal and, based on the echo, determines the proximity of a specific object. The frequency is outside the human audible range, so you won't hear a thing. Ultrasonic sensors can measure the distance of an object accurately at a minimum of 2 centimeters and a maximum of 3 meters from the device.

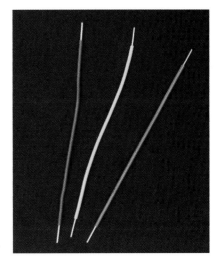

Figure 3-7. *Jumper wires*

This type of sensor functions well when you need to know not just whether something exists in front of an object, but also at what proximity. Lights don't affect ultrasonic sensors, so the sensors can function in complete darkness. On the other hand, there is a chance the sensor won't detect reflective surfaces or objects located at steep angles, because the sound wave won't bounce back from them. Also, very small or soft objects might reflect back such minuscule amounts of the sound that the sensor will not detect them.

The Parallax Ping))) Ultrasonic Sensor is shown in Figure 3-8.

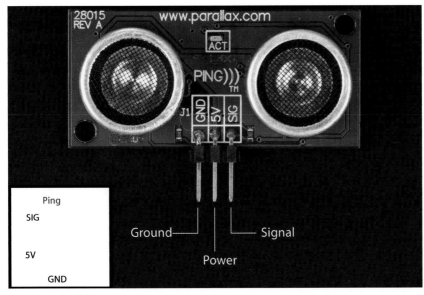

Figure 3-8. *The PING))) ultrasonic sensor*

METRIC UNITS

Because this book started life as a Finnish book, and because Arduino is an international phenomenon, we stuck with metric (also known as Système International, or SI) measurements throughout.

For easy conversions, you can type any measurement into a Google search, along with the desired output unit, such as "23 cm in inches," and the conversion calculation will appear in the results. In fact, you can get fancy, such as "29.112 microseconds per centimeter in mph," for which Google returns "768.389768 mph" (which is in fact roughly the speed of sound at 20 degrees C).

For a wider variety of calculations, check out Wolfram Alpha (*http://www.wolframalpha.com/*), which can not only tell you "speed of sound at 20 degrees Celsius," but will also let you tweak the results to take into account air pressure and humidity!

One negative aspect of ultrasonic sensors is their relatively high price ($30 at the time of this writing).

Measuring Distance with the Ping Ultrasonic Sensor

To begin creating the program, first connect the Arduino and upload the Blink code shown in Chapter 2. This brief exercise is always useful when you're starting a new project, because it confirms that the development environment still functions properly.

Insert the Arduino Nano in the middle of the prototyping breadboard, as shown in Figure 3-9. Pressing the Nano in all the way might require a bit of force, so be careful not to bend its pins. Next, connect the PING))) Ultrasonic Sensor to the Arduino by inserting it into the prototyping breadboard and connecting the Arduino pins to its pins with jumper wires.

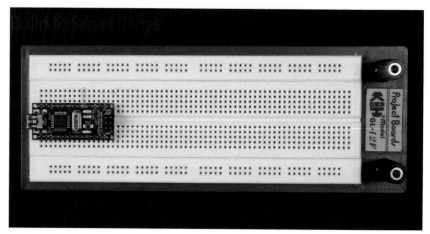

Figure 3-9. *Arduino Nano on a breadboard*

As shown in Figure 3-10, assignments for the three pins under the sensor are marked on the front of the sensor board. Remember, GND is an acronym for *ground*, or 0V. When a circuit includes a GND symbol, it is the same as connecting that part of the circuit to the negative terminal of the power supply. All grounds of a circuit are connected to one another and have the same voltage potential. Voltage always refers to a difference in electrical potential between two parts of a circuit. If only one voltage is given—for example, the +5V of the positive terminal of a power source such as a battery, or (in this case) an Arduino powered from USB—this voltage is compared to ground (GND), the negative terminal of the power source. The sensor's GND pin will be connected to the Arduino's GND pin. The sensor's 5V pin will receive power, and it is connected to the 5V pin of the Arduino.

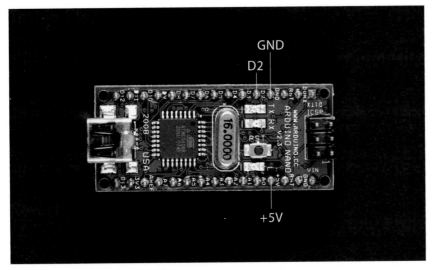

Figure 3-10. *Arduino Nano pin assignments*

The last leg of the sensor is marked with the abbreviation SIG (for *signal*). This pin will be connected to the Arduino digital pin 2 (D2). Through this connection, the Arduino will receive the data sent by the sensor and control it.

Insert the PING))) ultrasonic sensor into the prototyping breadboard close to the Arduino (see Figure 3-11), but don't let the Arduino and PING))) overlap any pins. Since the holes in the prototyping board are separated vertically, you must attach the pins of the PING))) sensor to the breadboard horizontally, each in its own vertical row.

Now the Arduino and PING))) sensors are connected to the prototyping breadboard, but they're not connected to each other yet. Let's do that now:

1. Connect the Arduino ground pin (GND) and PING))) ground pin to each other.

   To do this, connect one end of a black jumper wire to the same vertical row as the Arduino's GND pin. Connect the other end of the black jumper wire to the same vertical row as the PING))) GND pin.

2. Connect the Ping sensor's positive pin (5V) and the Arduino +5V to each other.

   To do this, connect one end of a red jumper wire to the same vertical row with the Arduino +5V pin. Connect the other end of the red jumper wire to the same vertical row as the Ping sensor's +5V pin.

3. Connect the PING))) SIG (signal) data pin and the Arduino digital pin D2 to each other.

   To do this, connect one end of a green (or blue, or yellow—anything but red or black) jumper wire to the same vertical row with the Ping SIG pin. Connect the other end of the green jumper wire to the same vertical row as the Arduino D2 pin.

Be careful when you connect wires to the Nano, because the pin assignments are printed so close together that it is easy to misread them. Each pin's label is printed above it, except the upper pins, which have their assignments written next to them. As of the Arduino Nano 3.0 release, all labels are printed adjacent to the pins and are much easier to read.

Figure 3-11 shows how the breadboard should appear; Figure 3-12 shows this circuit's schematic.

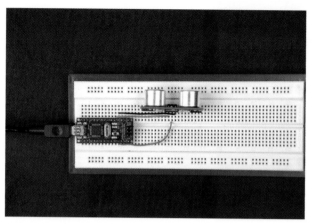

Figure 3-11. *A connected ultrasonic sensor*

Figure 3-12. *Schematic for the ultrasonic sensor circuit*

Distance-Measuring Program

Open the Arduino development environment and connect the Arduino to the computer's USB port. Feed the following code to Arduino by saving it into a new sketch and clicking Upload. This code is based on the Ping example that comes with the Arduino IDE (File→Examples→6. Sensors→Ping).

```
/* Ping))) Sensor

This sketch reads a PING))) ultrasonic rangefinder and returns the
distance to the closest object in range. To do this, it sends a pulse
to the sensor to initiate a reading, then listens for a pulse
to return. The length of the returning pulse is proportional to
the distance of the object from the sensor.

The circuit:
    * +V connection of the PING))) attached to +5V
    * GND connection of the PING))) attached to ground
    * SIG connection of the PING))) attached to digital pin 2

http://www.arduino.cc/en/Tutorial/Ping

created 3 Nov 2008
by David A. Mellis
modified 30 Jun 2009
by Tom Igoe
modified Nov 2010
by Joe Saavedra

*/

const int pingPin = 2;  ❶
long duration, distanceInches, distanceCm;  ❷
```

```
void setup()
{
  Serial.begin(9600);  ❸
}

void loop()
{
  pinMode(pingPin, OUTPUT);  ❹
  digitalWrite(pingPin, LOW);
  delayMicroseconds(2);
  digitalWrite(pingPin, HIGH);
  delayMicroseconds(5);
  digitalWrite(pingPin, LOW);

  pinMode(pingPin, INPUT);  ❺
  duration = pulseIn(pingPin, HIGH);

  distanceInches = microsecondsToInches(duration);  ❻
  distanceCm = microsecondsToCentimeters(duration);
  Serial.print(distanceInches);  ❼
  Serial.print("in, ");
  Serial.print(distanceCm);
  Serial.print("cm");
  Serial.println();

  delay(100);
}

long microsecondsToInches(long microseconds)
{
  return microseconds / 74 / 2;  ❽
}

long microsecondsToCentimeters(long microseconds)
{
  return microseconds / 29 / 2;  ❾
}
```

Let's look at what the code does:

❶ This constant won't change. It's the pin number of the sensor's output.

❷ Establish variables for the duration of the ping and the distance result in inches and centimeters.

❸ Initialize serial communication at 9,600 bits per second.

❹ The PING))) is triggered by a HIGH pulse of 2 or more microseconds. Give a short LOW pulse beforehand to ensure a clean HIGH pulse.

❺ The same pin is used to read the signal from the PING))): a HIGH pulse whose duration is the time (in microseconds) from when the ping is sent to when its echo off an object is received.

❻ Convert the time into a distance.

❼ Print the calculations for inches and centimeters to the Serial Monitor.

❽ According to Parallax's datasheet for the PING))), there are 73.746 microseconds per inch (i.e., sound travels at 1,130 feet per second). This gives the distance traveled by the ping, outbound, and return, so we divide by

2 to get the distance of the obstacle. The next function, `microseconds ToCentimeters()`, is explained next. It performs this calculation using metric (also known as SI) units, which is used worldwide.

❾ The speed of sound is 340 m/s or 29 microseconds per centimeter. The ping travels out and back, so to find the distance of the object we take half of the distance traveled. See *http://www.parallax.com/dl/docs/prod/ acc/28015-PING-v1.3.pdf.*

ADJUSTING FOR AIR TEMPERATURE

The PING))) example included with Arduino includes the number 29 in its `microsecondsToCentimeters()` function. Where does this number come from? Let's take a look:

The speed of sound in meters per second (*m/s*) at a given air temperature t (degrees Celsius) is calculated with this formula:

331.5 + 0.6 * t

With a temperature of 20° C, that's:

331.5 + 0.6 * 20 = 331.5 + 12 = 343.5 m/s

Let's convert speed to microseconds per centimeters. Start by converting to centimeters per second:

343.5 *100 = 34350 cm/s

In microseconds (*µs*), that's:

34350 / 1000000 = 0.03435 cm/us

Speed can be expressed as a pace, that is, how much time it takes to travel a given distance. Let's convert speed to pace, us/cm:

1/0.03435 = 29.112

If you plan to use this sketch in a location where the air temperature is much different than 20° C, you should perform these calculations with the appropriate air temperature. Note that if you intend to use the `microsecondsTo Inches()` function, you will need to convert your results accordingly.

Once you have uploaded the code successfully, the green light in front of the ultrasonic sensor should start blinking. Click the Arduino development environment's Serial Monitor button (shown in Figure 3-13). Now the distance values measured by the sensor should appear in the console on the bottom of the page. Move your hand in front of the sensor and observe how the values change.

```
000                    ping | Arduino 0021
⊳ ⊗  ▢ ⬆ ⬇ ⬌ ▣ Serial Monitor
 ping                                                    ⇨
const int pingPin = 2;
long duration, distanceInches, distanceCm;

void setup() {

  Serial.begin(9600);
}

void loop() {

  pinMode(pingPin, OUTPUT);
  digitalWrite(pingPin, LOW);
  delayMicroseconds(2);
  digitalWrite(pingPin, HIGH);
  delayMicroseconds(5);
  digitalWrite(pingPin, LOW);
```

Figure 3-13. *The Arduino development environment's Serial Monitor button*

Vibration Motor

A *vibration motor*, shown in Figure 3-14, is an ordinary DC (direct current) motor with an asymmetric weight. Asymmetrical rotation of the motor makes it vibrate. Think of the last time you had to balance a load of laundry in a washing machine: as long as that wet towel is stuck to one side of the drum, the washing machine is going to shake. The vibration motor works the same way.

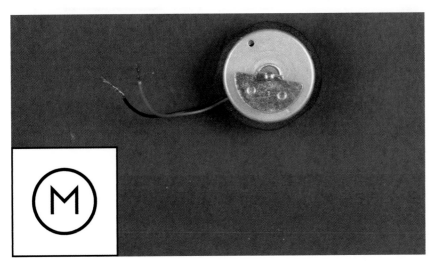

Figure 3-14. *Vibration motor*

You can find vibration motors at electronics stores (see "Parts," earlier in this chapter) or salvage one from an old device, such as a cell phone or video game controller with force feedback. The motor in Figure 3-14 came from a PlayStation controller. It is also helpful to keep the frame the motor was attached to. Not all vibration motors have a visible rotating part; in some, the part is packed inside a casing.

Connect the red motor wire to the Arduino digital pin number 5 (D5). Connect the black motor wire to the GND terminal row of the Arduino. If the wire is a multistrand (*stranded*) wire, it might fit too loosely in the prototyping bread-board. You can create a more secure attachment by using pliers to stick a small piece of metal wire into the same breadboard holes as the motor wire. Alternatively, you can solder the stranded wire to a piece of 22AWG solid-core wire and insert this into the breadboard. Figure 3-15 shows the vibration motor connected to the Arduino, and Figure 3-16 shows the schematic.

Figure 3-15. *Connecting a vibration motor*

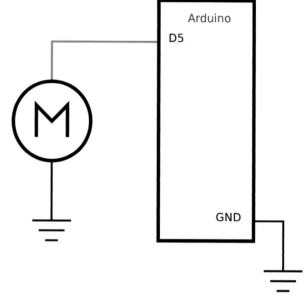

Figure 3-16. *Vibration motor schematic*

You can get the motor to turn by uploading the following code into the Arduino:

```
// dcMotor.pde - Run DC-motor to one direction
// (c) Kimmo Karvinen & Tero Karvinen http://botbook.com

int motorPin = 5; ❶

void setup() ❷
{
    pinMode(motorPin, OUTPUT); ❸
    digitalWrite(motorPin, HIGH); ❹
}

void loop() ❺
{
    delay(100);
}
```

When the motor receives enough power, it will start rotating. Let's go through the code:

❶ Set the value of the `motorPin` variable to 5. This makes the code easier to read because, instead of using a number, you can use the `motorPin` variable in certain parts of the code.

❷ Execute the `setup()` function only once, in the beginning. This function does not return any value; therefore, its type is empty (`void`).

❸ Switch pin D5 to an `OUTPUT` state. An Arduino pin can be set either as an output or an input. Because you would like to send power to the motor, choose `OUTPUT`.

❹ Turn on power to the D5 pin. The pin remains `HIGH` unless we set it to `LOW`, which makes the motor rotate continuously.

❺ The tasks your sketch performs will be written within the `loop()` function. An ordinary Arduino sketch spends most of its time repeating a loop. Since there is nothing to do in the `loop()` function, it is empty.

Combining Components to Make the Stalker Guard

Now you know how the necessary components work by themselves. Next, you will combine them to create the Stalker Guard.

First, connect the circuits for the motor and the ultrasonic sensor. Figure 3-17 shows them connected on the breadboard, and Figure 3-18 shows the schematic. Once you've connected them, it is wise to retry the code tested earlier for both devices, so you are testing only a single component at a time. This way, you ensure that you built the circuit correctly, which also makes troubleshooting easier in later stages.

If you can't find a vibration motor, you can use an LED as an indicator; replace the motor with an LED and a 220Ω resistor. (The Greek omega, Ω, is the symbol for ohm, the unit of resistance.)

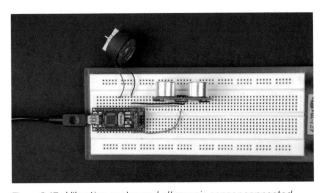

Figure 3-17. *Vibration motor and ultrasonic sensor connected*

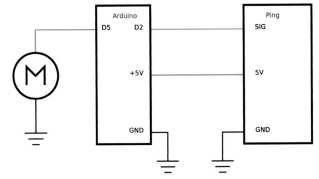

Figure 3-18. *Vibration motor and ultrasonic sensor schematic*

Making the Motor Vibrate

Now you will combine both of the sketches. To find out when a target is close, we'll make the vibration motor react when the distance to a target is below the specified limit:

```
/*
stalkerguard.pde - Shake if something comes near
(c) Kimmo Karvinen & Tero Karvinen http://botbook.com
updated 2010 - Joe Saavedra; based on code by David A. Mellis and Tom Igoe
*/
```

```
const int pingPin = 2;
const int motorPin = 5;
long int duration, distanceInches, distanceCm;
int limitCm = 60;       ❶

void setup()
{
  pinMode(motorPin, OUTPUT);
}

void loop()
{
  pinMode(pingPin, OUTPUT);
  digitalWrite(pingPin, LOW);
  delayMicroseconds(2);
  digitalWrite(pingPin, HIGH);
  delayMicroseconds(5);
  digitalWrite(pingPin, LOW);

  pinMode(pingPin, INPUT);
  duration = pulseIn(pingPin, HIGH);

  distanceInches = microsecondsToInches(duration);
  distanceCm = microsecondsToCentimeters(duration);

  checkLimit(); ❷
  delay(100);
}

void checkLimit()
{
  if (distanceCm < limitCm){   ❸
    digitalWrite(motorPin, HIGH);
  } else {        ❹
    digitalWrite(motorPin, LOW);
  }
}

long microsecondsToInches(long microseconds)
{
  return microseconds / 74 / 2;
}

long microsecondsToCentimeters(long microseconds)
{
  return microseconds / 29 / 2;
}
```

This program combines the earlier ultrasonic sensor and vibration motor code to determine whether the readings from the ultrasonic sensor are less than the defined threshold distance. If the distance is shorter, the motor will switch on.

❶ Add this variable to define the distance; when something gets this close, we will start the motor.

❷ Now that we have our current distance calculated, call the checkLimit function.

❸ If the distance is shorter than the defined threshold distance, switch on power to the motor pin.

❹   Otherwise, switch off the motor pin.

Providing Power from a Battery

Wearing the Stalker Guard when it's connected to a USB cable is a bit difficult, but you can free the Arduino from cables by attaching a battery to power it.

After uploading your code to the board, attach the positive (red) wire of the battery clip to the Arduino voltage in pin (VIN) terminal row of the prototyping breadboard. Similarly, attach the negative (black) wire to any Arduino GND pin. See Figures 3-19 and 3-20. Now the device will function via battery without the USB cable power. (When running off an external power source, the Nano needs at least 7V, but you should not give it more than 12V.)

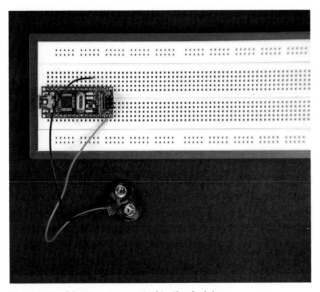

Figure 3-19. *A battery connected to the Arduino*

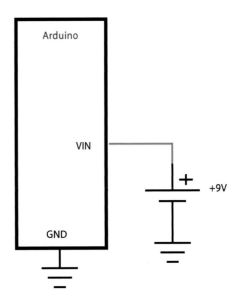

Figure 3-20. *Arduino battery connection schematic*

What's Next?

Now you have built your first prototype with the Arduino. Next, we will present one possible way to create an enclosure for the Stalker Guard. The proposed enclosure methods will give you suggestions on how to construct devices until the prototyping phase. Devices functioning on a prototyping breadboard do not necessarily make a major impression, except maybe on the most die-hard geeks. If your fingers are already itching to start building the next project, you can skip the enclosure stage and move on to the Robot Insect in Chapter 4.

Ideas for Future Projects

In the meantime, you can apply the skills you have learned to other projects. Here are a few ideas:

- Posture Watchdog: a device that warns you if you are leaning too close to your computer screen

- A distance-measuring device

- A robot that vibrates when you get close to it (see Figure 3-21)

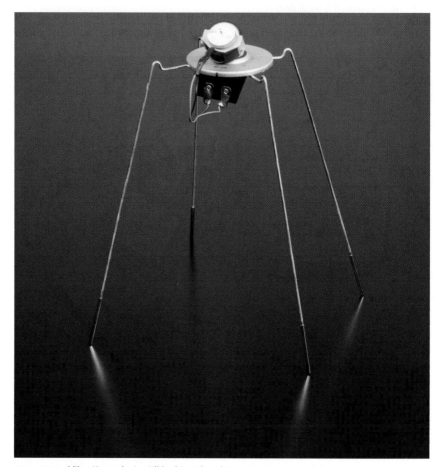

Figure 3-21. *Vibrating robot, still lacking the ultrasonic sensor*

Making an Enclosure

In the following sections, we'll outline one way to enclose the Stalker Guard to attach it to your waist and use it without wires. You don't need to follow these instructions exactly. Use them as a guideline and supply your own creativity and available parts.

Utilizing an Ammo Pouch

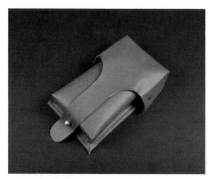

Figure 3-22. *The ammo pouch used as the basis for our enclosure*

As a starting point, we took an old ammo pouch purchased from an army surplus store (Figure 3-22). The pouch is made of rubber-coated linen fabric, which is relatively easy to perforate and cut. And its price—one euro—did not damage our budget.

Start by punching a hole for the screw, which will be used to attach the vibration motor cradle (Figure 3-23). If you do not have a ready-made cradle for the vibration motor, you can hot-glue it to the enclosure. The easiest way to create the hole is to use an electric drill, but you can also use a punch tool.

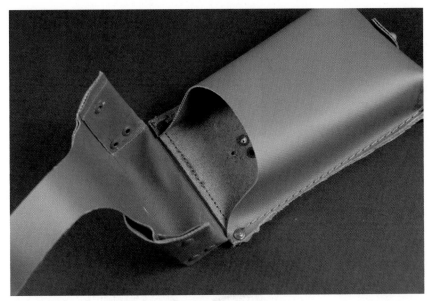

Figure 3-23. *Hole created for attaching the vibration motor*

Screw the cradle in place and attach the motor to it, as shown in Figures 3-24 and 3-25.

Figure 3-24. *The motor cradle in place*

Figure 3-25. *The vibration motor in place*

Mark the position for the ultrasonic sensor on top of the enclosure (Figure 3-26). You can create the holes with a mini-drill sanding bit, as shown in Figure 3-27. If you don't have a sanding bit, you can use a carpet knife.

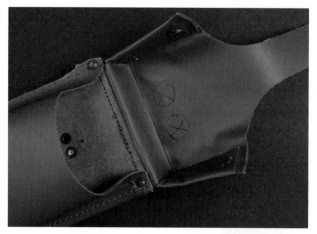

Figure 3-26. *The marked positions for the ultrasonic sensor*

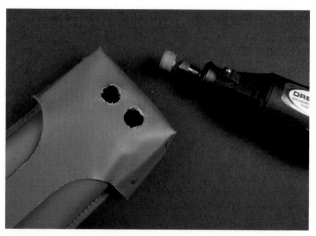

Figure 3-27. *Holes created with a mini drill*

Hot-glue the ultrasonic sensor in place, as shown in Figures 3-28 and 3-29. If you would like to be able to detach it later for other uses, you could use duct tape instead of glue.

Figure 3-28. *A hot-glued ultrasonic sensor*

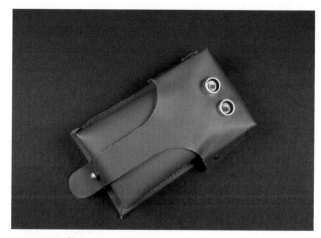

Figure 3-29. *The ultrasonic sensor in place*

Wiring Up the Circuit

The circuits presented earlier in this chapter were made with a small prototyping breadboard. At this stage, you could also consider soldering the device into one single package. However, we did not want to permanently attach the Arduino and the ultrasonic sensor to the Stalker Guard, so we went with a simpler solution: a servo extension cable (Figure 3-30), which has a female connector suitable for attaching to the sensor pins at each end. Cut off the connector from one end (Figure 3-31) and solder single-strand jumper wires to the wire, as shown in the next section, "Soldering Basics." By using an extension cable, you can attach the sensor tightly without destroying it.

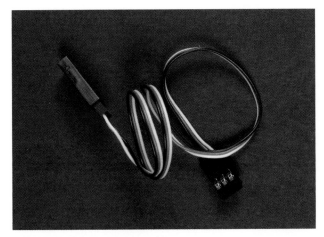

Figure 3-30. *Servo extension cable*

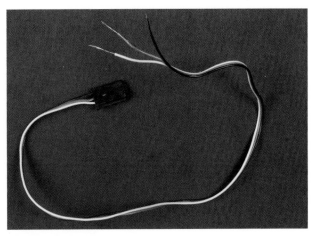

Figure 3-31. *Connector removed from one end of the servo extension cable*

You should also solder single-strand jumper wires to the vibration motor to make it stay connected better.

Soldering Basics

The projects in this book require a few simple solderings, such as attaching wires to each other and to the battery clip. While doing the exercises, you'll get more experience with soldering, and possible mistakes won't end up causing major problems like broken components. As with any tool, you must use eye protection when working with solder.

Regardless of what you are soldering, always try to position parts so that they are in place as securely as possible. This makes working so much easier. If you are connecting two wires to each other, twist together the ends to be soldered, as shown in Figures 3-32 and 3-33. If you're using stranded wire, twist the strands together first. It is harder to solder strands that point in all directions.

The Make: Electronics Deluxe Toolkit includes everything you need to get started, including hand tools, soldering iron, soldering stand and sponge, as well as wire, solder, and a multimeter: http://www.makershed.com/ProductDetails.asp?ProductCode=MKEE1.

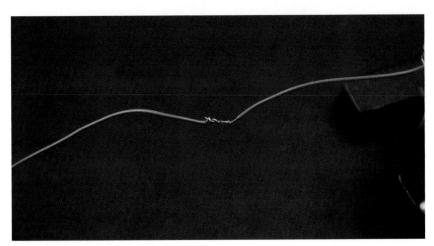

Figure 3-32. *Wires with ends twisted together*

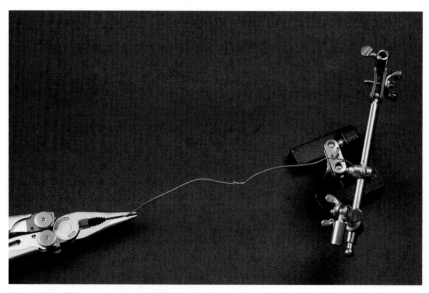

Figure 3-33. *Wires supported solidly by a holder*

Swipe the hot soldering iron with a wet sponge to get rid of any excess solder, as shown in Figure 3-34. Do not flick solder off the iron.

Figure 3-34. *Swiping the soldering iron with a wet sponge*

Heat the parts to be soldered *quickly*—no more than one second (Figure 3-35). Heating for too long can damage parts that are not meant to be heated, destroying sensitive components or melting plastic parts. It is good to practice with wire first.

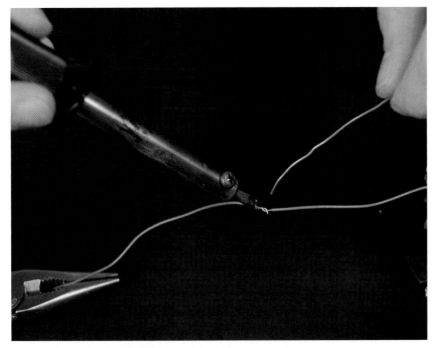

Figure 3-35. *Heating the parts for approximately one second*

With the soldering iron still touching the parts to be soldered, push the solder so that a suitable amount of it flows in (Figure 3-36); this should take one second. Take the solder away from the joint and then remove the soldering iron as well (Figure 3-37). When you have had enough practice, the whole process will take only two or three seconds.

Do not touch the wire or component that you are soldering. It can get hot enough to hurt or give you a small burn.

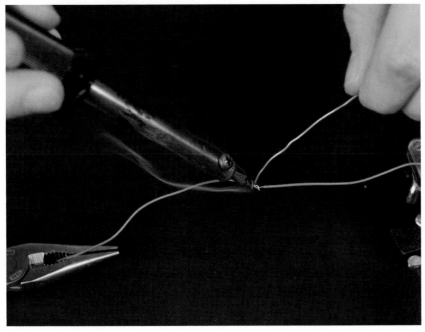

Figure 3-36. *Adding solder*

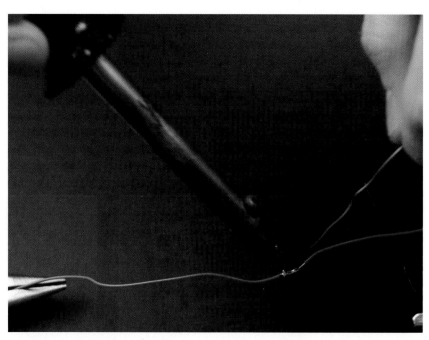

Figure 3-37. *Removing the soldering iron and solder from the joint*

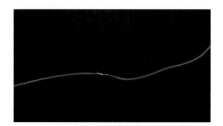

Figure 3-38. *Finished soldered joint*

It is important to heat the parts to be soldered to a high enough temperature to melt the solder. If hot solder is just dropped on the wires to be soldered, you will create a cold solder joint. Cold solder joints might not conduct electricity properly and will probably fail. Figure 3-38 shows the finished solder joint. Figure 3-39 shows wires attached to the servo extension cable.

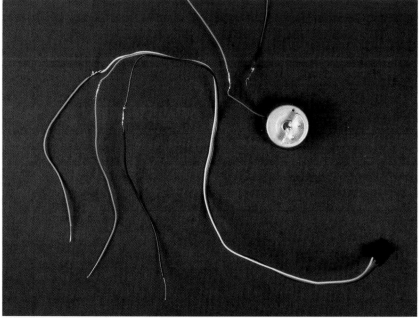

Figure 3-39. *Single-strand wires soldered to the servo extension cable and to the vibration motor*

Chapter 3

Soldering might be a bit uncomfortable at first, but just like with all other mechanical tasks, practice makes perfect.

Using a Switch to Save Batteries

This section shows you how to install a switch between the negative cable of the battery clip and the Arduino, so that the Arduino won't consume batteries while the device is not in use (Figure 3-40). This way, you won't have to open the enclosure to operate the Stalker Guard; you can just turn it on and off with the switch.

Connect the positive (red) wire of the battery clip to the Arduino +5V pin in the prototyping breadboard. Connect the negative (black) wire from the battery clip to one pin of the switch. Connect the other pin of the switch to the Arduino GND pin. Now Arduino gets its power from the battery when the switch is in the correct position. Drill a hole in the enclosure for screwing in the switch.

Figure 3-40. *A switch attached to the enclosure*

It is helpful to place *heat-shrink tubing* (rubber tubing that will shrink in diameter by 50% when heated) on the soldered wires (Figure 3-41) to avoid shorts inside the enclosure. Put a piece of heat-shrink tubing over the area and heat it with a heat gun or hair dryer, as shown in Figure 3-42. Be sure to do this in an area with adequate ventilation.

If your heat gun uses an external flame, be careful to not overheat the tubing or you may melt the insulation of the wires or damage nearby components.

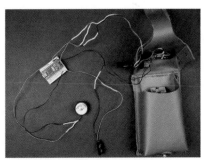

Figure 3-41. *All components attached together; heat-shrink tubing covers the soldered points*

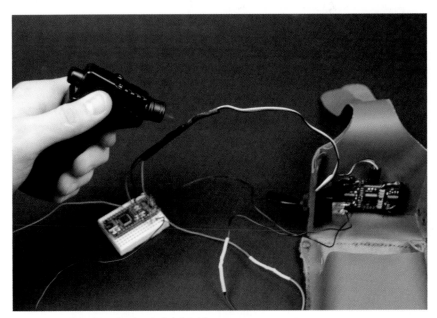

Figure 3-42. *Attaching the heat-shrink tubing*

You can place larger heat-shrink tubes over smaller ones to keep the wires neatly together, as shown in Figure 3-43.

Figure 3-43. *All components attached together; larger heat-shrink tubing neatly covers the individual soldered points covered by smaller heat-shrink tubing*

> *You can purchase heat-shrink tubing from electronic suppliers, including many RadioShack stores. If you don't have any heat-shrink tubing immediately available, electrical or insulating tape can also get the job done, but don't use a heat gun on the tape.*

Figure 3-44 shows the Stalker Guard schematic, Figure 3-45 shows the finished Stalker Guard, and Figure 3-46 shows the device in use.

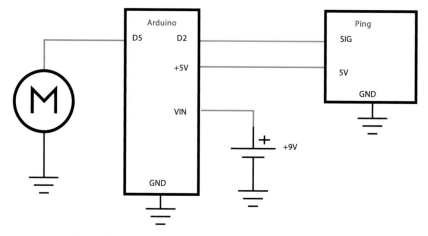

Figure 3-44. *Connection diagram*

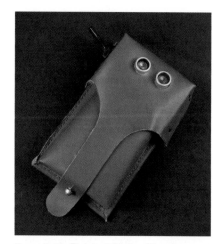

Figure 3-45. *Finished Stalker Guard*

Figure 3-46. *The Stalker Guard in use*

Insect Robot

4.

In this project, you'll create an Insect Robot (see Figure 4-1) that walks forward on four legs. Using ultrasound, the robot can see in the dark, just like a bat. When it detects an obstacle, it takes a few steps back, turns, and continues forward. The robot can walk over small obstructions, and it looks more human than its wheeled relatives. Once you've learned the techniques in this chapter, you can easily extend your Insect Robot with new tentacles and sensors—and new code.

Before starting this project, you should know what ultrasonic sensors are, and make sure that the "Hello World" Blink code from Chapter 2 is working properly with your Arduino. You'll learn about some new things, including *servo motors* (motors that can be manipulated to rotate to a specific angular position), in this chapter.

We will build a body for the insect by gluing two servos together and shaping legs for it from a wire clothes hanger. Arduino will turn the two servos one at a time, which moves each pair of metal legs like real legs so the insect can crawl forward. We will also make holders for the battery and Arduino so our insect can behave autonomously.

The insect will need eyes to react to its environment. We will connect an ultrasonic sensor to the insect's head to enable the robot to precisely measure its distance from objects in front of it.

Finally, we will teach the insect to react to an obstacle by backing up and turning. The distance of the obstruction will trigger a series of commands that make the robot back up several steps, turn, and move forward several more steps. Then our Insect Robot will be able to move and avoid obstructions on its own.

After you have spent some time observing your new pet's movements, you can add new sensors to the robot or teach it new tricks.

Figure 4-1. *The completed Insect Robot*

What You'll Learn

In this chapter, you'll learn:

- How to control servos

- The basics of robot walking

- How to construct mechanical structures for prototypes

- How to build a walking robot based on two servos

Tools and Parts

You'll need the following tools and parts for this project (Figure 4-2).

Manufacturer part numbers are shown for:

- *Maker SHED (US: http://maker-shed.com): SHED*

- *Element14 (International and US; formerly Farnell and Newark, http://element-14.com): EL14*

- *SparkFun (US: http://sparkfun.com): SFE*

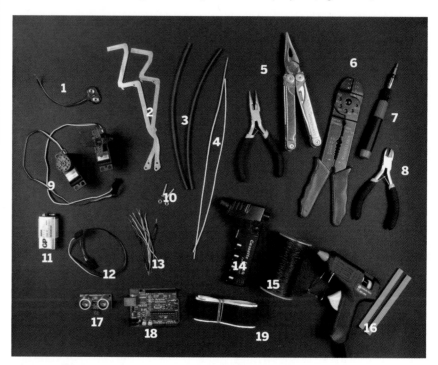

Figure 4-2. *The tools and parts you need to build the insect*

1. 9V battery clip (EL14: 34M2183; SFE: PRT-00091).

2. Two small metal rods. You could salvage these from other devices, such as an old typewriter. If you have metal snips and a small amount of sheet metal, you could also cut them yourself (but be sure to use a metal file or metal sandpaper to smooth the edges, which will be extremely sharp).

3. Heat-shrink tubing (14cm) for the feet (EL14: 90N7288). Hot glue works well, too.

4. 28cm and 25cm pieces from a wire clothes hanger.

5. Two pairs of pliers.

6. Wire strippers (EL14: 61M0803; SFE: TOL-08696).

7. Multihead screwdriver.

8. Diagonal cutter pliers (EL14: 52F9064; SFE: TOL-08794).

9. Two large servo motors (SFE: ROB-09064; *http://parallax.com/Store*: 900-00005).

10. Two nuts and bolts.

11. 9V battery.

12. Servo extension cable (SFE: ROB-08738; *http://parallax.com/Store*: 805-00002).

13. Red, black, and yellow (or some other color) jumper wire (SHED: MKEL1; EL14: 10R0134; SFE: PRT-00124).

14. A butane torch or a cigarette lighter.

15. Thin metal wire.

16. Hot-glue gun and hot glue (available at craft stores or office supply stores).

17. PING))) ultrasonic sensor (SHED: MKPX5; *http://www.parallax.com/Store/*).

18. Arduino Uno (SHED: MKSP4; EL14: 13T9285; SFE: DEV-09950). The older Duemilanove is pictured here and would work just as well.

19. Hook-and-loop fastener tape, such as Velcro.

Servo Motors

Servo motors (Figure 4-3) come in different sizes and prices and are based on different technologies. In this context, we are talking about *hobby* servos, the kind used in remote-control cars, for example. Servo motors have a servo controller that directs the position of the motor wherever we want.

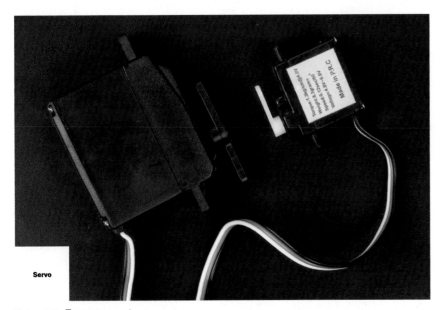

Figure 4-3. *Two servo motors*

The motor itself is a DC (direct current) motor with gears. Servo motors usually rotate rather slowly and with a relatively strong torque.

You can buy hobby servo motors with either *limited* rotation or *continuous* rotation. Limited rotation models work for most purposes, and you can control their movement quite precisely by degrees of rotation. In continuous rotation servos, you can control only speed and direction. We'll cover continuous rotation servos in Chapter 8.

Wiring Up the Circuit

> *The colors can vary depending on the motor, but generally, the darkest color is GND, red is power, and the next lightest color is the data wire.*

Connect the servo to the Arduino by attaching the servo's black wire to any of the Arduino GND pins. The red wire indicates positive voltage; connect it to the Arduino +5V pin.

The white or yellow data wire controls the servo; you will connect it to one of the digital pins. For this project, connect the data wire to the first available digital pin: digital pin 2 (D2). Figure 4-4 shows the connection, and Figure 4-5 shows the schematic.

> *Arduino uses D0 and D1 to communicate over USB or serial, so it's best to use those to connect things to your project only when absolutely necessary.*

Figure 4-4. *Servo connection*

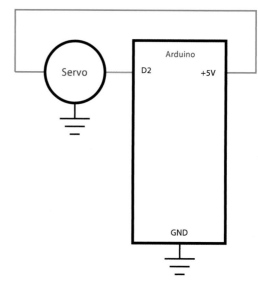

Figure 4-5. *Arduino/servo schematic*

Using the Servo Library in Arduino

Libraries are collections of subroutines or classes that let us extend the basic functionality of a platform or language such as Arduino. There are many different libraries that help us interpret data or use specific hardware in much simpler and cleaner ways. You can explore the libraries available for Arduino at *http://arduino.cc/en/Reference/Libraries*. As these libraries are meant to extend our code only when needed, we must declare each library in any sketch where one will be used. We do this with a single line of code. Here's how to include the library for controlling servo motors:

```
#include <Servo.h>
```

Now we can reference methods and objects from within that library at any time in our sketch.

We will be using the Servo library to interface with our motors in this chapter. The Servo library comes with a standard installation of Arduino and can support up to 12 motors on most Arduino boards and 48 motors on the Arduino Mega. For each servo motor we are using, we must create an instance of the Servo object with Servo myServo;.

In the setup() function, we must associate this instance of Servo to a specific pin, the same pin to which the data wire of our motor is attached, using the command myServo.attach(2);.

Now talking to our motor is easy. There are several functions for communicating with it, including read(), write(), detach(), and more, all of which you can explore in the library reference at *http://arduino.cc/en/Reference/Servo*. For this chapter, when talking to our motors, we will use only the write() function, which requires a single argument: degree of rotation.

Centering the Servo

This example shows how we use the Servo library to connect a single servo motor and rotate it toward the absolute center point.

```
// servoCenter.pde - Center servo
// (c) Kimmo Karvinen & Tero Karvinen http://BotBook.com
// updated - Joe Saavedra, 2010
#include <Servo.h>  ❶

Servo myServo;  ❷

void setup()
{
  myServo.attach(2);  ❸
  myServo.write(90);  ❹
}

void loop()
{
  delay(100);
}
```

❶ Import the library.

❷ Create an instance of Servo and name it myServo.

❸ Attach myServo to pin 2.

❹ Tell the servo to rotate 90 degrees.

Because we want the motor to move to one position and stay there, we can include all our code in the setup() function. The loop() function must be declared for Arduino to compile, but because we don't need to do anything in the loop, it can remain empty.

When we write() to a servo, we set it to a specific position. Limited rotation servo motors can turn from 0 to 180 degrees, so setting ours to 90 turns it exactly half of its maximum rotation. The servo is now perfectly centered, and it will remain that way until given further instruction.

Moving the Servo

Let's write a small program that will rotate the servo first to the center, then to the maximum angle, back to center, and then to the minimum angle.

```
// moveServo.pde - Move servo to center, maximum angle
// and to minumum angle
// (c) Kimmo Karvinen & Tero Karvinen http://BotBook.com
// updated - Joe Saavedra, 2010
#include <Servo.h>

Servo myServo;
int delayTime = 1000;  ❶

void setup()
{
  myServo.attach(2);
}

void loop()
{
  myServo.write(90);  ❷
  delay(delayTime);  ❸

  myServo.write(180);  ❷
  delay(delayTime);

  myServo.write(90);
  delay(delayTime);
  myServo.write(0);
  delay(delayTime);
}
```

These are the only differences between this and the previous servo code:

❶ The variable delayTime variable (set to 1,000 milliseconds, or 1 second) determines how much time to wait between each rotation.

❷ Set the servo to a new angle of rotation, and then wait for a specified duration of time.

❸ The duration, named delayTime, must also take into account how quickly the motor can turn. In this case, we must wait a minimum of 375 milliseconds before sending the motor a new command. It takes this much time for the motor to rotate 90 degrees. Play with the value of this variable. You will notice that any value less than 375ms is not enough time for the motor to reach its destination, so it will begin to malfunction.

❹ Similarly, you can rotate the servo to other positions simply by changing the values written to myServo. Any value between 0 and 180 will function properly, because this is our range of rotation. In this example, these values are hardcoded, meaning they are written explicitly on each line. In the future, we'll store these values in variables for more complicated and efficient applications.

Now you can rotate servos. Which objects would you like to move in your own embedded systems?

Constructing the Frame

Now it's time to build the frame for the robot.

Making the Legs

Cut two pieces from a wire clothes hanger: 28cm for the rear legs and 25cm for the front legs, as shown in Figure 4-6. Bend the legs with pliers, as shown in Figure 4-7. It's important to make the legs long enough and to make sure that the feet point backward, which lets them act as hooks and enables the robot to climb over obstacles. At this stage, don't worry too much about the shape of the legs. You can adjust them later (and will likely need to).

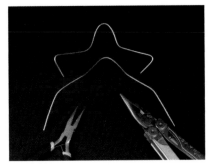

Figure 4-7. *Front legs (top) and rear legs (bottom)*

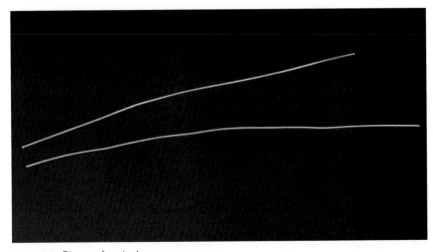

Figure 4-6. *Pieces of a wire hanger*

The legs will have a better grip if you cover them with *heat-shrink tubing*, as shown in Figure 4-8. Heat-shrink tubing is rubber tubing that will shrink in diameter by 50% when heated, for example, with a heat gun or hair dryer. Cut two 7cm pieces of the tubing and shrink them to fit around the back legs, as shown in Figure 4-9.

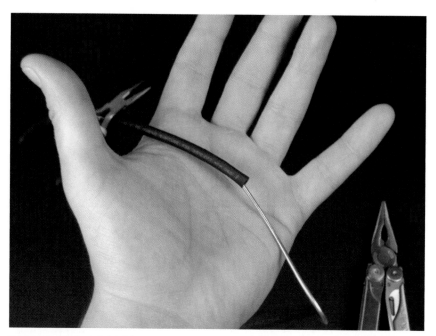

Figure 4-8. *Heat-shrink tubing*

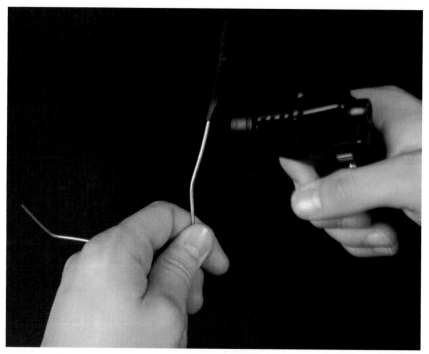

Figure 4-9. *Attaching the heat-shrink tubing*

Next, attach the legs to the servos. The servos come with one or more plastic attachments that will connect to the servo axis. Attach the legs by pulling metal wires through the servo holes, and secure each leg by tightening the metal wire, as shown in Figure 4-10. Cut any excess wire to keep it from hindering the motor's movement. Finally, add hot glue to the underside to stabilize the legs, as shown in Figure 4-11, but do not fill the center screw hole with glue.

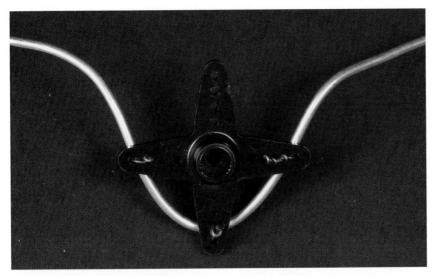

Figure 4-10. *Attaching with metal wire*

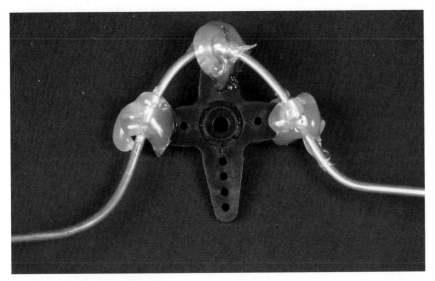

Figure 4-11. *Securing with hot glue*

Assembling the Frame

The frame of the walker consists of two connected servo motors. Before gluing them together, you need to remove a small plastic extension (meant for mounting the servos) from both of the servo motors. Remove the extension next to the servo arm from the front-facing servo, and remove the opposite part from the rear-facing servo. You can do this easily with a utility knife, as shown in Figure 4-12. It's a good idea to smooth out the cutting seam with a small file to make sure that the glued joints do not become uneven and weak.

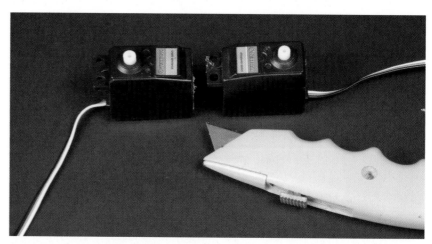

Figure 4-12. *Removing the obstructing plastic*

Spread the hot glue evenly on the rear servo, as shown in Figure 4-13, and immediately press the servos together (Figure 4-14), holding them steady for a while to give the glue time to set. The servos are connected so that the front-facing servo arm points forward and the rear-facing arm points down. The top sides of the motors should be placed evenly, to make it easier to attach them to the Arduino. If you make a mistake gluing the parts together, it's easy to separate them without too much force. (Hot-gluing is not necessarily ideal for building sturdy devices, but it is a quick and easy way to attach almost anything, and it works well with simple prototypes.)

Figure 4-13. *Spreading the hot glue*

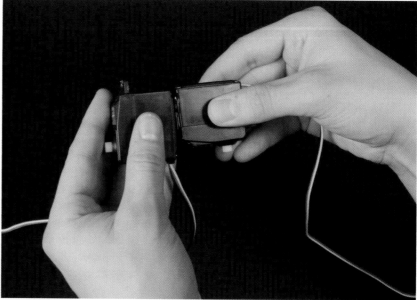

Figure 4-14. *Gluing the servos together*

Chapter 4

Making the Holder for the Arduino

We will use two metal strips to build a holder on top of the robot that will make it easy to attach and detach the Arduino. Cut two 10cm metal pieces. Bend the sides of each strip so that the space in the middle is equal to the width of the Arduino (Figure 4-15) and glue both strips to the servos, as shown in Figure 4-16.

Figure 4-15. *Bent attachments*

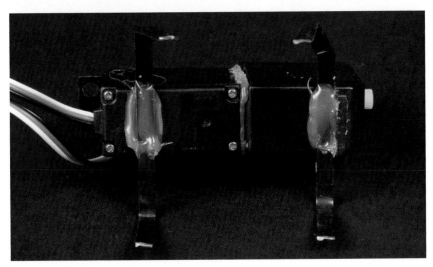

Figure 4-16. *Gluing attachments*

The Arduino Duemilanove (and later models, such as the Uno) used in this project is 5.2cm wide. If the metal you're using is flexible enough, it is helpful to bend the corner inward slightly. This way, you can snap the Arduino in place sturdily and, when you are finished, remove it painlessly for other uses.

Attaching a Battery

We'll use Velcro tape to make an attachment system in the rear of the robot for the 9V battery. Cut a 16cm strip from the Velcro tape and attach the ends together. Make holes for two screws in the middle of the tape. Attach the Velcro tape with screws to the servo's extension part in the rear of the robot, as shown in Figure 4-17.

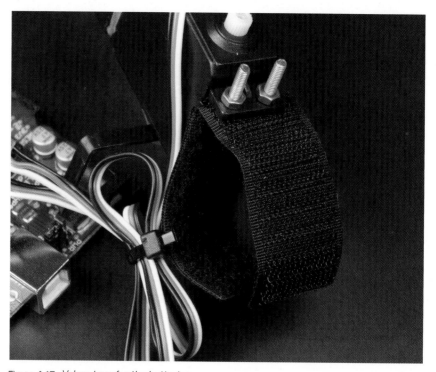

Figure 4-17. *Velcro tape for the batteries*

Assembly

Now you can place the Arduino board on top of the servos. Attach the legs to the servos, but don't screw them in tightly yet.

We'll connect the servos to each other and to the Arduino with jumper wires. First, connect the servos' black (GND) wires using a black jumper wire from one servo to the other, and then use another black jumper wire to connect one of the servos to an Arduino GND pin. You might have to use a bit of force to insert two wires into one of the servo connection headers.

Next, connect the red power wires in a similar manner, first from one servo to the other and then to the Arduino's 5V power pin. Use white (the actual color may vary) jumper wires to control each servo, and connect a yellow jumper wire from the rear servo to Arduino pin 2 and from the front servo to Arduino pin 3.

Figure 4-18 shows the connections being made. Note that only one of the yellow wires is connected; you'll need to connect one yellow wire to each servo. The schematic (Figure 4-19) shows this in more detail.

Figure 4-18. *Connections*

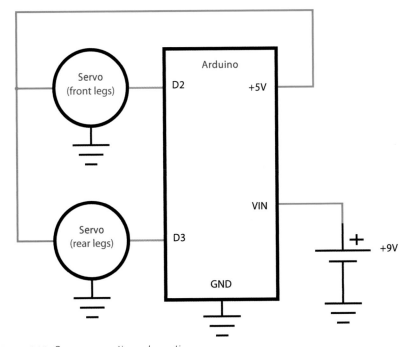

Figure 4-19. *Servo connection schematic*

Earlier in the chapter, in the "Centering the Servo" section, you ran a sketch to center a single servo. If you run this code again to center the servo, you will be able to attach the leg in the correct position. But there are now two servos, so let's alter the previous centering code to turn both servos toward the center:

```
//twoServosCenter.pde - Center two servos
// (c) Kimmo Karvinen & Tero Karvinen http://BotBook.com
//updated - Joe Saavedra, 2010
#include <Servo.h>

Servo frontServo;
Servo rearServo; ❶

void setup()
{
  frontServo.attach(2);
  rearServo.attach(3); ❷
  frontServo.write(90); ❸
  rearServo.write(90);
}

void loop()
{
    delay(100);
}
```

The only difference between this and the earlier centering code is the addition of two servo objects named frontServo and rearServo:

❶ Define an instance of the Servo object for the rear servo.

❷ Within setup, attach the rearServo to pin 3.

❸ Send pulses to both of the motors, making them turn toward the center.

Screwing the Legs in Place

Now that the servos are centered, you can screw the legs into place. It's also a good idea to attach the battery now, because the additional weight will affect the way the robot walks. However, it is not necessary to connect the battery wires to the Arduino; you can take power straight from the USB cable while you are programming and testing the device. Figure 4-20 shows the finished robot.

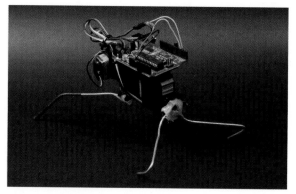

Figure 4-20. *The finished robot frame*

Wires rarely fit perfectly tight to Arduino ports or servo extension cables. Constantly disconnecting wires makes building painful. To attach a wire more securely, bend its end into a small curve, as shown in Figure 4-21.

If you really want to make sure that the wires stay where they should, use the ScrewShield. The ScrewShield adds "wings" with terminal blocks to both sides of Arduino. Terminal blocks have screws so you can attach one or more wires firmly to any pin. SHED: MKWS1, SFE DEV-09729, *http://adafruit.com*: 196, and on *http://store.fungizmos.com*.

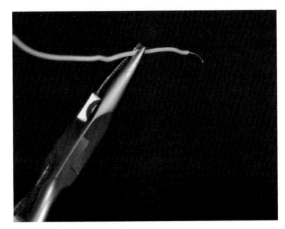

Figure 4-21. *Bending the wire*

Figure 4-22. *The ScrewShield*

Programming the Walk

Now we can program the Arduino to walk.

Walking Forward

If you power up the Arduino running the code that swings only one servo (see the earlier "Moving the Servo" section), it will start rocking on either its front or rear legs. Walking forward will require coordination between both front and rear legs. When the servos move at the same tempo, but in opposite directions, the robot starts to walk. Here is some code that will make the robot walk forward:

```
// walkerForward.pde - Two servo walker. Forward.
// (c) Kimmo Karvinen & Tero Karvinen http://BotBook.com
// updated - Joe Saavedra, 2010
#include <Servo.h>

Servo frontServo;
Servo rearServo;
int centerPos = 90; ❶
int frontRightUp = 72; ❷
int frontLeftUp = 108; ❸
int backRightForward = 75; ❹
int backLeftForward = 105;❺
void moveForward() ❻
{
    frontServo.write(frontRightUp);
```

```
    rearServo.write(backLeftForward);
    delay(125);
    frontServo.write(centerPos);
    rearServo.write(centerPos);
    delay(65);
    frontServo.write(frontLeftUp);
    rearServo.write(backRightForward);
    delay(125);

    frontServo.write(centerPos);
    rearServo.write(centerPos);
    delay(65);
}

void setup()
{
    frontServo.attach(2);
    rearServo.attach(3);
}

void loop()
{
    moveForward();  ❼
    delay(150);  //time between each step taken, speed of walk
}
```

Let's have a look at the code:

❶ This is the center position for the servos. Ninety degrees is precisely half of 180 possible degrees of rotation.

❷ Maximum position the right front leg will rise to.

❸ Maximum position the left front leg will rise to.

❹ Maximum position the right rear leg will bend to.

❺ Maximum position the left rear leg will bend to.

❻ The moveForward function turns the servos first to opposite directions. The variables defined in the preceding lines set how far each of the servos will rotate. Before we turn in another direction, we will tell the servos to rotate toward a predefined center point for a short span of time. This ensures that the servos don't start drifting out of sync. We return to the center point at the end of every step to make the walk more elegant and efficient.

❼ Call the moveForward function repeatedly within the loop, which will make our robot move one step forward. The subsequent delay controls how long the robot waits before taking its next step. Removing the delay is the equivalent of having the robot run as fast as it can.

Walking Backward

When walking forward works, walking backward is easy. This time, the servos move at the same pace and in the same direction:

```
// walkerBackward.pde - Two servo walker. Backward.
// (c) Kimmo Karvinen & Tero Karvinen http://BotBook.com
// updated - Joe Saavedra, 2010
```

```
#include <Servo.h>

Servo frontServo;
Servo rearServo;
int centerPos = 90;
int frontRightUp = 72;
int frontLeftUp = 108;
int backRightForward = 75;
int backLeftForward = 105;

void moveBackward()  ❶
{
  frontServo.write(frontRightUp);
  rearServo.write(backRightForward);
  delay(125);
  frontServo.write(centerPos);
  rearServo.write(centerPos);
  delay(65);
  frontServo.write(frontLeftUp);
  rearServo.write(backLeftForward);
  delay(125);
  frontServo.write(centerPos);
  rearServo.write(centerPos);
  delay(65);
}

void setup()
{
  frontServo.attach(2);
  rearServo.attach(3);
}

void loop()
{
  moveBackward();  ❷
  delay(150);  //time between each step taken, speed of walk
}
```

Let's see what has changed from the previous code:

❶ The moveBackward() function is similar to moveForward(), but this time the right front leg will rise up when the right rear leg moves forward, and the left front leg rises when the left rear leg moves forward.

❷ Now moveBackward() is called in the loop() function.

Turning Backward

Moving the robot forward and backward is not enough if we want it to avoid obstacles. The preferred outcome is for the robot to detect the obstacle, turn in another direction, and continue to walk. Naturally, it could just back up and turn after that, but the turn would be more efficient if the robot first backs up to the right and then turns to the left.

The robot can turn to the right as it walks backward if you alter the center point of the servo and the threshold levels a bit toward the right side. This will also change the balance of the robot, which can easily lead to one of its front legs rising higher than the other. You can solve this problem by adding a bit of movement to the lowered leg, raising it into the air:

```
// walkerTurnBackward.pde - Two servo walker. Turn backward.
// (c) Kimmo Karvinen & Tero Karvinen http://BotBook.com
// updated - Joe Saavedra, 2010
#include <Servo.h>

Servo frontServo;
Servo rearServo;
int centerPos = 90;
int frontRightUp = 72;
int frontLeftUp = 108;
int backRightForward = 75;
int backLeftForward = 105;

void moveBackRight()  ❶
{
  frontServo.write(frontRightUp);
  rearServo.write(backRightForward-6);
  delay(125);
  frontServo.write(centerPos);
  rearServo.write(centerPos-6);
  delay(65);
  frontServo.write(frontLeftUp+9);
  rearServo.write(backLeftForward-6);
  delay(125);
  frontServo.write(centerPos);
  rearServo.write(centerPos);
  delay(65);
}

void setup()
{
  frontServo.attach(2);
  rearServo.attach(3);
}

void loop()
{
  moveBackRight();
  delay(150);  //time between each step taken, speed of walk
}
```

The new `moveBackRight()` function is similar to the `moveBack()` function in the previous example.

❶ The movement of the rear servo is reduced by 6 degrees, which will move its center point to the right. As we noted earlier, the changed rear servo position will likely change the balance of the entire robot. To account for this, we add 9 degrees to the `frontLeftUp` value. If any of your own robot's legs stay in the air or drag, you can increase or decrease these values as needed.

Turning Forward

A turn forward resembles otherwise normal forward walking, but now the center points of both servos are changed. The movement of one front leg must also be adjusted to keep it from rising too high.

```
// walkerTurnForward.pde - Two servo walker. Turn forward.
// (c) Kimmo Karvinen & Tero Karvinen http://BotBook.com
// updated - Joe Saavedra, 2010
#include <Servo.h>

Servo frontServo;
Servo rearServo;
int centerTurnPos = 81; ❶
int frontTurnRightUp = 63; ❷
int frontTurnLeftUp = 117; ❸
int backTurnRightForward = 66; ❹
int backTurnLeftForward = 96;❺

void moveTurnLeft() ❻
{
  frontServo.write(frontTurnRightUp);
  rearServo.write(backTurnLeftForward);
  delay(125);
  frontServo.write(centerTurnPos);
  rearServo.write(centerTurnPos);
  delay(65);
  frontServo.write(frontTurnLeftUp);
  rearServo.write(backTurnRightForward);
  delay(125);
  frontServo.write(centerTurnPos);
  rearServo.write(centerTurnPos);
  delay(65);
}

void setup()
{
  frontServo.attach(2);
  rearServo.attach(3);
}

void loop()
{
  moveTurnLeft(); ❼
  delay(150);  //time between each step taken, speed of walk
}
```

❶ Create a new variable that will define the center point of the servos during a turn (the *center turn position*). Notice we are 9 degrees away from the motor's halfway point.

❷ Calculate the maximum position to which the right front leg will rise by deducting 18 degrees from the center turn position.

❸ Calculate the position of the left front leg in the same way as the right front leg, but instead of adding 18 degrees to the center turn position, add 36 (to balance the front legs).

❹ Calculate the position for the right rear leg by deducting 15 degrees from the center turn position.

⑤ Add the same distance to the left rear leg.

⑥ The MoveTurnLeft() function is similar to all the other walks. This time, the variables just mentioned are used for turning, and both servos are centered to a different position than when walking forward.

⑦ Repeat turning forward in the loop and the delay to control the speed of our walk.

Now you have made your robot walk in different directions and turn while moving forward and backward. Next, we will combine all the movements into one sketch and coordinate movement with an ultrasonic range sensor.

If you have any problems with how your robot walks, it is best to adjust only one walking direction at a time.

Avoiding Obstacles Using Ultrasound

Now it's time to add an ultrasonic sensor to our robot.

Attaching the Ultrasonic Sensor

Use the servo extension cable to attach the sensor. You can remove the plastic part covering the cables from one end to expose the wires, as shown in Figure 4-23. If you'd like, you could cut the connector off and solder three single-strand wires as you did with the Stalker Guard enclosure in Chapter 3.

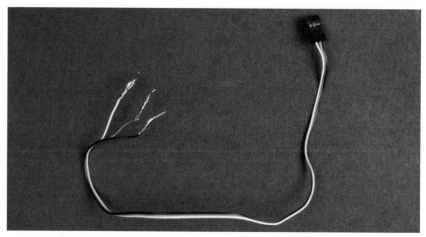

Figure 4-23. *One end of the servo extension cable disassembled*

Using hot glue, we connected the other end of the extension cable to the front of the robot with the connector part pointing down (if you'd rather have it pointing up, that will work, too). If the connector doesn't snap in easily, you can glue the wires in place. The pin assignments are marked on top of the pins. After the glue has dried, you can snap the ultrasonic sensor into place on the connector.

Connect the other end of the servo extension cable to the Arduino. The red goes to the Arduino 5V port, black to the GND port, and white to the digital pin 4. Since the Arduino does not have many free pins left, you can connect the black and red wires in the same holes as the servo cables. Figure 4-24 shows the sensor connected to the robot (the schematic is shown in Figure 4-25).

Chapter 4

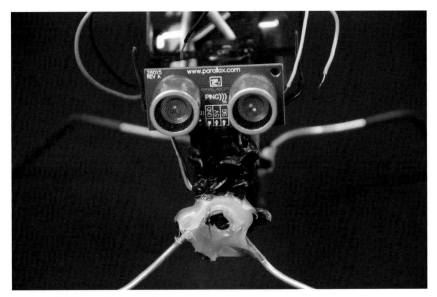

Figure 4-24. *Ultrasonic sensor in place*

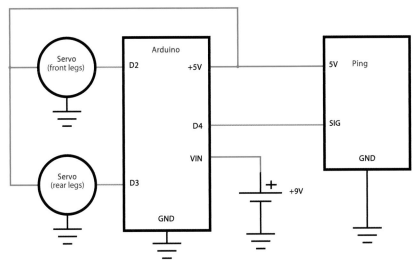

Figure 4-25. *Ultrasonic sensor and servo connection schematic*

Code

Next, we'll connect all previously tested movements into one program. In addition, we'll add some ultrasonic code for detecting obstacles (see Chapter 3 for the Stalker Guard project, which uses similar code).

```
// walkerForwardComplete.pde - Two servo walker.
// Complete code with obstacle avoidance
// (c) Kimmo Karvinen & Tero Karvinen http://BotBook.com
// Updated by Joe Saavedra, 2010
#include <Servo.h>

Servo frontServo;
Servo rearServo;
```

```
/* Servo motors - global variables */ ❶
int centerPos = 90;
int frontRightUp = 72;
int frontLeftUp = 108;
int backRightForward = 75;
int backLeftForward = 105;
int walkSpeed = 150; // How long to wait between steps in milliseconds
int centerTurnPos = 81;
int frontTurnRightUp = 63;
int frontTurnLeftUp = 117;
int backTurnRightForward = 66;
int backTurnLeftForward = 96;

/* Ping distance measurement - global variables */
int pingPin = 4;
long int duration, distanceInches;
long distanceFront=0; //cm
int startAvoidanceDistance=20; //cm

long microsecondsToInches(long microseconds)
{
  return microseconds / 74 / 2;
}

long microsecondsToCentimeters(long microseconds)
{
  return microseconds / 29 / 2;
}

long distanceCm(){ ❷
  pinMode(pingPin, OUTPUT);
  digitalWrite(pingPin, LOW);
  delayMicroseconds(2);
  digitalWrite(pingPin, HIGH);
  delayMicroseconds(5);
  digitalWrite(pingPin, LOW);

  pinMode(pingPin, INPUT);
  duration = pulseIn(pingPin, HIGH);

  distanceInches = microsecondsToInches(duration);
  return microsecondsToCentimeters(duration);
}

void center()
{
  frontServo.write(centerPos);
  rearServo.write(centerPos);
}

void moveForward() ❸
{
  frontServo.write(frontRightUp);
  rearServo.write(backLeftForward);
  delay(125);
  frontServo.write(centerPos);
  rearServo.write(centerPos);
  delay(65);
```

Chapter 4

```
  frontServo.write(frontLeftUp);
  rearServo.write(backRightForward);
  delay(125);

  frontServo.write(centerPos);
  rearServo.write(centerPos);
  delay(65);
}

void moveBackRight()
{
  frontServo.write(frontRightUp);
  rearServo.write(backRightForward-6);
  delay(125);
  frontServo.write(centerPos);
  rearServo.write(centerPos-6);
  delay(65);
  frontServo.write(frontLeftUp+9);
  rearServo.write(backLeftForward-6);
  delay(125);

  frontServo.write(centerPos);
  rearServo.write(centerPos);
  delay(65);
}

void moveTurnLeft()
{
  frontServo.write(frontTurnRightUp);
  rearServo.write(backTurnLeftForward);
  delay(125);
  frontServo.write(centerTurnPos);
  rearServo.write(centerTurnPos);
  delay(65);
  frontServo.write(frontTurnLeftUp);
  rearServo.write(backTurnRightForward);
  delay(125);

  frontServo.write(centerTurnPos);
  rearServo.write(centerTurnPos);
  delay(65);
}

void setup() ❹
{
  frontServo.attach(2);
  rearServo.attach(3);
  pinMode(pingPin, OUTPUT);
}

void loop() ❺
{
  distanceFront=distanceCm(); ❻
  if (distanceFront > 1){ // Filters out any stray 0.00 error readings ❼
    if (distanceFront<startAvoidanceDistance) { ❽
      for(int i=0; i<=8; i++) { ❾
        moveBackRight();
        delay(walkSpeed);
      }
```

```
      for(int i=0; i<=10; i++) {
        moveTurnLeft();
        delay(walkSpeed);
      }
    } else {
      moveForward();  ❿
      delay(walkSpeed);
    }
  }
}
```

Because we went through the code for various turning techniques earlier, we will cover only the combination of the sketches here:

❶ Declare global variables in the beginning, gathering all global variables here from previous sketches. At the same time, check that there are no conflicts with the names. For example, centerPos is from the walker code, and t is from the ping distance sensor code in Chapter 3.

❷ These functions (microsecondsToInches, microsecondsToCentimeters, and distanceCm) come from the distance sensor code in Chapter 3.

❸ Pull in functions relating to walking such as moveForward() from earlier sketches in this chapter.

❹ Add all the lines from the setup() function within the new program's setup() function. The setup code for the pin modes is identical to the previous example. The setting of the pingPin state and declaration, as well as the setting of variables v and speedCmUs, is copied from the distance meter code in Chapter 3.

❺ There is new code in the main program's loop() function. This is the central program logic.

❻ Measure a distance toward an object in front of us with the distanceCm() function.

❼ Sometimes, the PING))) sensor might return an incorrect reading of 0.00. This is not uncommon for sensors of this type; however, we must compensate for these false readings with a simple *filter*. This if statement allows only readings above 1cm to pass through.

❽ If the measured distance (distanceFront) is longer than 1cm but shorter than the declared startAvoidanceDistance, an obstacle is detected and must be avoided.

❾ Avoid the obstacle by backing up and turning to the right for nine steps. The delay for walkSpeed is keeping the rhythm of our steps consistent. Then, take 11 steps forward and turn simultaneously to the left.

❿ If there is no obstacle within the 20cm range, take a step forward with moveForward().

Now the insect can walk forward. It will also avoid obstacles without touching them.

What's Next?

You now have a new pet (Figure 4-26), and you can control servos. You have learned techniques related to construction, such as hot-gluing and using heat-shrink tubing. You have become familiar with the challenges of walking within the field of robotics. You can use all these techniques in your own projects.

Ideas for the Next Stage

- Add new sensors to your insect. For example, teach it to move toward the light or to shy away from movement.

- Add a sensor to the front, enabling the robot to detect an obstacle and raise its legs higher.

- Create a code for the ultrasonic sensor that detects within meaningful intervals whether any movement has taken place. This way, you can make sure your robot has not gotten stuck anywhere.

- Attach a piezo speaker to the robot to make sounds and provide information about the robot's operations.

- Teach your insect to stand up if it falls on its back.

Figure 4-26. *The Insect Robot walking in an autumn forest*

Interactive Painting

You wave your hand in front of a painting, and a new image slides in. As this chapter explains, this kind of "interactive painting" is made possible by sensors that control computer programs. In this project, you will learn how to connect Python and Processing animations with Arduino to create an interactive slideshow (see Figure 5-1) you control with your hands.

In this project, ultrasonic sensors connected to Arduino will follow the movement of a user's hand as he waves it in front of a display. Arduino will forward information about the movements to the computer serial port. Python or Processing (we'll provide examples in both languages) will then execute the commands indicated by the specific hand motion, moving the images on the display accordingly.

The purpose of this chapter is to show you how to use data sent by the Arduino in formats other than just the development environment's console window. You'll be able to apply this knowledge easily to projects such as motion-controlled games or theft alarms controlled by a computer. You'll learn that your projects don't need to be limited to using just a keyboard and a mouse; you can supplement those traditional tools with sensors that measure almost anything you'd like—such as light, temperature, motion, or pressure.

You will learn how to use resistors and LEDs during this project. Before we start, however, you should know how an ultrasonic sensor operates (Chapter 3 explains this in detail).

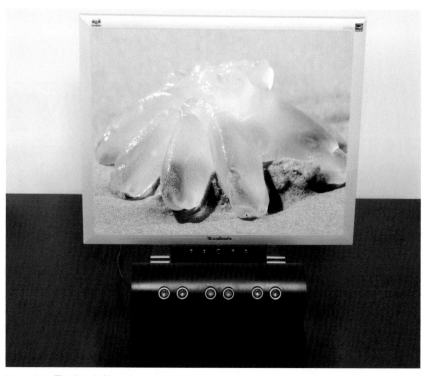

Figure 5-1. *The finished project*

What You'll Learn

In this chapter, you'll learn how to:

- Limit current flow with a resistor and read resistor color codes
- Measure resistance with a multimeter
- Use LEDs as a light signal
- Install Python or Processing
- Program a simple graphical user interface
- Read information from Arduino to the computer using Python and Processing

Tools and Parts

In this project, you'll need the following tools and parts (shown in Figure 5-2).

Manufacturer part numbers are shown for:
- *Maker SHED (US: http://makershed.com): SHED*
- *Element14 (International and US; formerly Farnell and Newark, http://element-14.com): EL14*
- *SparkFun (US: http://sparkfun.com): SFE*

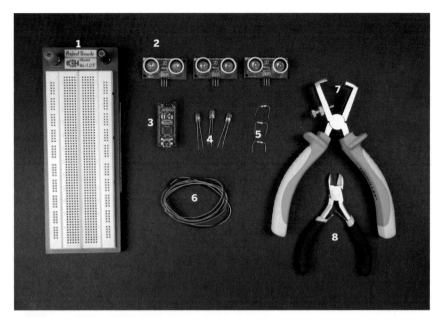

Figure 5-2. *Parts used in this project*

1. Solderless breadboard (SHED: MKEL3; SFE: PRT-00112; EL14: 15R8319).

2. Three PING))) ultrasonic sensors (SHED: MKPX5; *http://www.parallax.com/ Store/*).

3. Arduino Nano (SHED: MKGR1; *http://store.gravitech.us*; or *http://store .gravitech.us/distributors.html*).

4. Three green LEDs (EL14: 40K0064; SFE: COM-09592).

5. Three resistors between 220 and 330 ohms (EL14: 58K5042). Purchase individually or as part of a resistor assortment.

6. Jumper wires in three colors (SHED: MKEL1; EL14: 10R0134; SFE: PRT-00124).

7. Wire strippers (EL14: 61M0803; SFE: TOL-08696).

8. Diagonal cutter pliers (EL14: 52F9064; SFE: TOL-08794).

Resistors

A *resistor* restricts the flow of electricity. We'll use resistors in this project to drop the 5V voltage on the Arduino's digital pins to a level safe enough to connect the LEDs. Without the resistors, the LEDs might burn out. Figure 5-3 shows a few resistors, along with two different resistor symbols you may encounter in a schematic. The top one is commonly used in the US, while the bottom one may be found elsewhere in the world (the bottom symbol is used in this book).

Resistance is measured in *ohms*. This value is marked on the side of a resistor using a color-coding scheme in which each color represents a number (see Figure 5-4). Projects in this book use resistors with four color rings. The code is

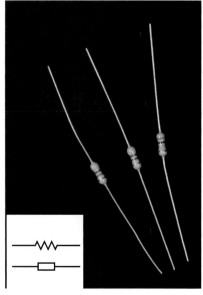

Figure 5-3. *Resistors*

	Black	0
	Brown	1
	Red	2
	Orange	3
	Yellow	4
	Green	5
	Blue	6
	Violet	7
	Gray	8
	White	9

Figure 5-4. *Meanings of resistor colors*

read from left to right, with the ring that is farthest from the others positioned on the right side. The first two rings are read as numbers, and the third tells us the number of zeros in our multiplier. The fourth ring marks the *tolerance*, a percentage that indicates the possible deviation from the resistor's rated value.

Hold the resistor so that the tolerance ring is on the right. In Figure 5-5, the first ring is orange and the second one is white (which represents 3 and 9, respectively). The third ring, which indicates the number of zeros in our multiplier, is brown (which represents 1 zero, so you'll multiply the other two numbers by 10). Resistance is therefore 39×0, or 390, ohms.

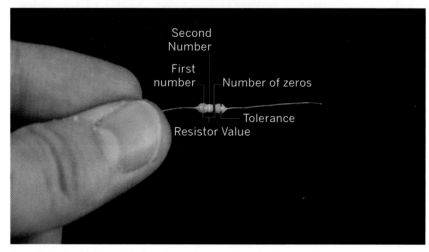

Figure 5-5. *Reading a resistor's value*

Measuring Resistance with a Multimeter

If you are unsure about the value you read from the color rings, you can double-check the result using a multimeter, as shown in Figure 5-6.

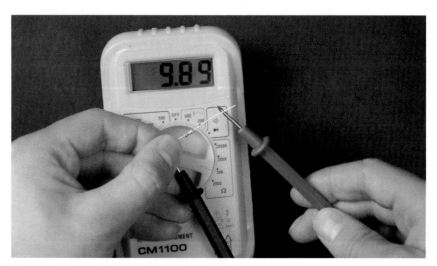

Figure 5-6. *Measuring resistance with a multimeter*

　　　　Chapter 5

Turn the multimeter to its lowest resistance range and firmly press the multimeter probes to the resistor. Resistors do not have polarity, which means you can position the black and red cables either way. You can use alligator clips to hold the probes in place.

If there is no number on the multimeter's screen, or if the number is –1, switch to a higher resistance range. Continue until you see a result on the screen.

When you get a reading, check the multiplier in your chosen resistor measuring range. If the reading is 47 and the measurement range is "20 kilohms" or "20kΩ," the result is therefore 47 kilohms, or 47,000 ohms.

For the lazier measurers, meters with an automatic range selection are available. You can find multimeters from suppliers such as SparkFun (*http://www.sparkfun.com/*) and Element 14 (*http://www.element-14.com/*).

> ## THIRD COLOR RING TRICK
>
> Usually, the precise value of the resistor does not matter so much, as long as it is close to the one you need. Normally, you can find a suitable resistor from a large assortment of them by using the following trick.
>
> Check the multiplier of the resistor. The multiplier is usually the third ring. For example, if you are searching for a few-hundred-ohm resistor to use in front of an LED, the multiplier is 1 (brown). So you should simply search for a resistor that has a brown third band.
>
> There are two numbers in front of the multiplier ring: the first and second rings. That is the reason why the multiplier is one fewer than the amount of zeros we need in our measured resistance value. For example, to find a resistor valued in thousands of ohms (three zeros), we will need a multiplier of 2 (red). For tens of thousands (4 zeros), we will need a multiplier of 3 (orange).

LEDs

An *LED*, or light-emitting diode, is a semiconductor that radiates light when an electrical current runs through it. LEDs (Figure 5-7) are often used in instrument lighting, as signal lights, and even in household lighting. Compared to traditional types of lamps, LEDs are smaller, mechanically sturdier, and longer lasting.

Infrared (IR) LEDs emit light at a wavelength longer than that of the visible spectrum. IR light is invisible to the eye.

Since an LED has a very short switch-on and shutoff response time, you can flicker it to transmit information. This application is used in many products, for example, remote controls. You can also use infrared LEDs to detect obstacles.

LEDs require only a small amount of current, so a resistor must usually be connected in front of it to prevent damage resulting from too much current. Choosing the right operating voltage will make an LED last longer.

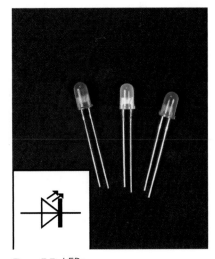

Figure 5-7. *LEDs*

Because an LED is in fact a diode, current in an LED will flow only in one direction. For this reason, it's important to note which leg is the positive (anode) and which is the negative (cathode, ground). One way is by observing leg length. The shorter leg is always the ground of the diode, and the longer is the positive leg. Another method is to look for a small, flattened spot in the plastic next to the shorter leg of an LED. This side must be connected to ground.

Now we'll connect an LED to the Arduino.

Insert the Arduino into the prototyping breadboard. The holes in the prototyping breadboard are connected in vertical rows, split by a stripe in the center of the board (the *gutter*). Position the Arduino so that it straddles the gutter evenly, as shown in Figure 5-8. This way, you can avoid shorting opposite Arduino pins.

Connect the longer (positive) leg of the LED on the same vertical row as the Arduino pin D2. Then put the shorter (negative) leg in one of the empty vertical rows and connect one end of the resistor (either one is fine, since resistors are not polarized) on that same row.

Connect the free end of the resistor to the same vertical row as the Arduino GND pin.

The circuit is now ready. You've connected the Arduino D2 pin via an LED and a resistor to ground. Do you recognize each component in the breadboard shown in Figure 5-8 and the connection diagram shown in Figure 5-9?

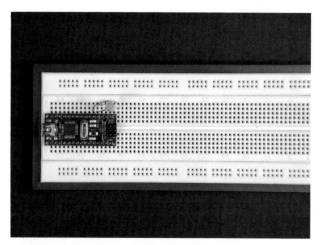

Figure 5-8. *Connecting an LED to the Arduino Nano*

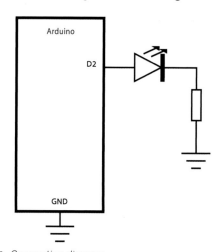

Figure 5-9. *Connection diagram*

The following code switches the LED on for half a second and then shuts it off for half a second.

```
// (c) Kimmo Karvinen & Tero Karvinen http://BotBook.com

int ledPin = 2; ❶

void setup()
{
    pinMode(ledPin, OUTPUT); ❷
}
```

```
void loop()
{
    digitalWrite(ledPin, HIGH); ❸
    delay(500); ❹
    digitalWrite(ledPin, LOW); ❺
    delay(500);
} ❻
```

Here's a look at the code:

❶ Store the pin number of the LED in the variable `ledPin`. This way, you do not have to change this value in all the other lines of code if you want to change the pin you are using later on.

❷ Set the pin to `OUTPUT` mode, so that the `digitalWrite()` command can control its state.

❸ Set the pin to `HIGH` mode, which means it will supply +5V power.

❹ Wait for 500 milliseconds, i.e., half a second.

❺ Set the pin to `LOW` mode (this means it is no longer supplying current).

❻ When the `loop()` ends, automatically return to the beginning of the `loop()` and repeat continuously.

Detecting Motion Using Ultrasonic Sensors

In this step, you'll connect three ping ultrasonic sensors to the Arduino. In addition to the three sensors and the Arduino, you need a prototyping breadboard and wire to build this circuit.

The ground (negative) conductor is the same all over the circuit. Therefore, connect the GND pins of all the ultrasonic sensors to the Arduino using a black jumper wire. Connect the +5V pins of the sensors to the Arduino +5V pin using red jumper wires. Our prototyping breadboard has long plus and minus rails on the side, so we connected the wires using these rails.

Finally, connect each ultrasonic sensor data pin to its own digital data pin in the Arduino, as shown in Figures 5-10 and 5-11. Connect the left sensor to pin D2, the middle sensor to pin D3, and the right sensor to pin D4. The holes in the prototyping breadboard are connected to one another in vertical rows. For example, connect the left sensor data wire directly to the hole in the breadboard above the Arduino D2 pin.

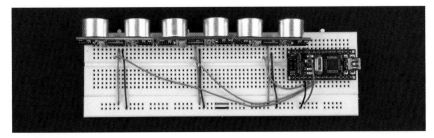

Figure 5-10. *The three ultrasonic sensors connected*

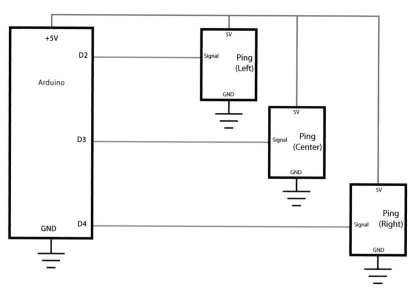

Figure 5-11. *Connection diagram*

Make sure that the circuit works. The easiest way to test it is with the Ping sensor Arduino sketch covered in the section "Ping Ultrasonic Sensor" in Chapter 3. You can find this example by clicking File, then choosing Examples→6→Sensors→Ping.

```
/* Ping))) Sensor

   This sketch reads a PING))) ultrasonic rangefinder and returns the
   distance to the closest object in range. To do this, it sends a pulse
   to the sensor to initiate a reading, then listens for a pulse
   to return. The length of the returning pulse is proportional to
   the distance of the object from the sensor.

   The circuit:
    * +V connection of the PING))) attached to +5V
    * GND connection of the PING))) attached to ground
    * SIG connection of the PING))) attached to digital pin 7

   http://www.arduino.cc/en/Tutorial/Ping

   created 3 Nov 2008
   by David A. Mellis
   modified 30 Jun 2009
   by Tom Igoe

 */

const int pingPin = 2;

void setup()
{
  Serial.begin(9600);
}
```

```
void loop()
{
  long duration, inches, cm;

  pinMode(pingPin, OUTPUT);
  digitalWrite(pingPin, LOW);
  delayMicroseconds(2);
  digitalWrite(pingPin, HIGH);
  delayMicroseconds(5);
  digitalWrite(pingPin, LOW);

  pinMode(pingPin, INPUT);
  duration = pulseIn(pingPin, HIGH);

  inches = microsecondsToInches(duration);
  cm = microsecondsToCentimeters(duration);

  Serial.print(inches);
  Serial.print("in, ");
  Serial.print(cm);
  Serial.print("cm");
  Serial.println();

  delay(100);
}

long microsecondsToInches(long microseconds)
{
  return microseconds / 74 / 2;
}

long microsecondsToCentimeters(long microseconds)
{
  return microseconds / 29 / 2;
}
```

Compile the code and upload it to the Arduino. Switch on the serial console from the Arduino development environment in the computer. Remember to set the same speed for the serial console—9,600 bits per second—that is used in the code.

For an explanation of the numbers used in microsecondsToCentime ters() *and* microsecondsToInches()*, see the sidebar titled "Adjusting for Air Temperature" in Chapter 3.*

When you wave your hand in front of an active ultrasonic sensor, the distance is displayed in the serial console in centimeters.

Double-check that the distance reading is accurate, at least within 10 centimeters. In some configurations, there's a glitch in the Arduino development environment in which the pulseIn() function gets stuck or returns total gibberish (values that are a hundred or a thousand times larger than they should be). If this happens, updating the Arduino development environment version to a newer one might help.

When you have made sure that the first sensor functions correctly, test the other sensors the same way. Change the code first:

```
int pingPin = 3;
```

Finally, test the third sensor by changing the code:

```
int pingPin = 4;
```

Now you know that you have three functional ultrasonic sensors and a working Arduino. You have also made sure that you can read the distance from an ultrasonic sensor.

Reading All the Sensors

Now, here's a program that reads the distance from all three sensors. To efficiently use this code, we will move the sensor measurement section of the code into its own function called `getDistance()`. This way, we do not have to worry about the details of measuring distance, but can instead use the code like a library. We only want the distance in centimeters, and that will be provided by the `getDistance()` function. All other sections—variables, declaration of functions, and contents of the `setup()`—are simply copied from the preceding code.

> Chapter 3 covers the operation of an ultrasonic sensor in detail.

```
/* Read 3 Ping Sensors
   Based on code by
   David A. Mellis and Tom Igoe
   Joe Saavedra, 2010
   http://BotBook.com
 */

const int leftPing = 2; ❶
const int centerPing = 3;
const int rightPing = 4;

void setup()
{
  Serial.begin(9600); // bit per second
}

void loop()
{
  Serial.print("Left: ");
  getDistance(leftPing); ❷
  Serial.print("Center: ");
  Serial.println(getDistance(centerPing));
  Serial.print("Right: ");
  Serial.println(getDistance(rightPing));
  Serial.println(); ❸

  delay(250); // ms ❹
}

int getDistance(int pingPin)
{
  long duration, inches, cm;

  pinMode(pingPin, OUTPUT);
  digitalWrite(pingPin, LOW);
  delayMicroseconds(2);
  digitalWrite(pingPin, HIGH);
  delayMicroseconds(5);
  digitalWrite(pingPin, LOW);

  pinMode(pingPin, INPUT);
  duration = pulseIn(pingPin, HIGH);
```

```
  inches = microsecondsToInches(duration);
  cm = microsecondsToCentimeters(duration);

  //return(inches);
  return(cm);  ❺
}

long microsecondsToInches(long microseconds)
{
  return microseconds / 74 / 2;
}

long microsecondsToCentimeters(long microseconds)
{
  return microseconds / 29 / 2;
}
```

This code is identical to the earlier example, except that what was previously inside the `loop()` is now in a function called `getDistance()`. For `get Distance()` to run, we must pass in an integer, `pingPin`, which is the pin that the function reads a value from. Now, we only have to call the `getDistance()` function three times, each time passing in a pin number for the appropriate sensor. Let's view the control of the left sensor in the order of execution:

❶ Store the pin numbers to the variables. The name of the variable provides the position of the sensor, so this naming convention makes the code much clearer than just using a plain number (such as 2).

❷ Retrieve the distance from the left pin (and then print it out on the following lines). Let's break this statement down:

```
getDistance(leftPing);
```

This line reads the distance from the left sensor. Since the value of the variable is 2, the call is equivalent to:

```
getDistance(2);
```

If a hand is, for example, 12 centimeters away from the sensor connected to pin 2, the function pulses that pin accordingly and returns the calculated distances.

❸ `Serial.println();` prints a blank line between readings. It is added simply to make the data being printed in the Serial Monitor easier to read.

❹ This `delay()` pauses the program for 250 milliseconds after every reading. This is exactly one fourth of a second. This pause slows down the rate of data capture, but is important in this code so that we can read the Serial Monitor clearly. Our final application will have a much shorter delay time for faster readings.

❺ These lines control what value is returned to the `loop()`. Currently, the inches calculation has been commented out, so only the centimeter calculation will be returned.

After you compile the code, be sure to open the Serial Monitor to watch data from all three sensors print out. Make sure each sensor reacts to your hand moving.

Testing the Circuit Using LEDs

Let's write a program that lights an LED when a hand moves closer than 20cm to a sensor. When the hand waves in front of the display, the LEDs near each of the three sensors will light up accordingly, one by one.

Add three LEDs to the circuit, the long (positive) leg of each LED connected to an Arduino data pin, and the shorter (negative) legs (next to the small indent) connected to the ground via a resistor. Connect the LEDs to pins D5, D6, and D7, as shown in Figures 5-12 and 5-13.

Figure 5-12. *LEDs connected for testing*

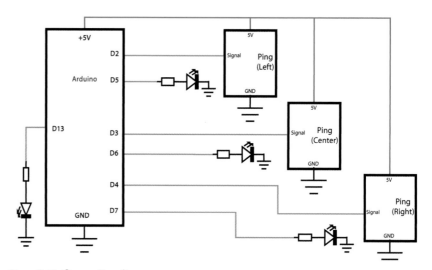

Figure 5-13. *Connection diagram*

```
/* Read 3 Ping Sensors
Based on code by
David A. Mellis, Tom Igoe,
Kimmo Karvinen and Tero Karvinen
Updated by Joe Saavedra, 2010
http://BotBook.com
*/

const int leftPing = 2;
const int centerPing = 3;
const int rightPing = 4;
```

```
const int leftLed = 5; ❶
const int centerLed = 6;
const int rightLed = 7;

const int maxD = 20; // cm; maximum hand distance ❷

void setup()
{
  Serial.begin(9600);
  pinMode(leftLed, OUTPUT); ❸
  pinMode(centerLed, OUTPUT);
  pinMode(rightLed, OUTPUT);
}

void loop()
{
  ping(leftPing, leftLed); ❹
  ping(centerPing, centerLed);
  ping(rightPing, rightLed);
  delay(50);
}

boolean ping(int pingPin, int ledPin) ❺
{
  int d = getDistance(pingPin); // cm ❻
  boolean pinActivated = false; ❼
  if (d < maxD) {    ❽
    digitalWrite(ledPin, HIGH);
    pinActivated = true;
  } else {
    digitalWrite(ledPin, LOW);
    pinActivated = false;
  }
  return pinActivated;
}

int getDistance(int pingPin)
{
  long duration, inches, cm;

  pinMode(pingPin, OUTPUT);
  digitalWrite(pingPin, LOW);
  delayMicroseconds(2);
  digitalWrite(pingPin, HIGH);
  delayMicroseconds(5);
  digitalWrite(pingPin, LOW);

  pinMode(pingPin, INPUT);
  duration = pulseIn(pingPin, HIGH);

  inches = microsecondsToInches(duration);
  cm = microsecondsToCentimeters(duration);

  return(cm); // You could also return inches
}

long microsecondsToInches(long `microseconds)
{
  return microseconds / 74 / 2;
}
```

```
long microsecondsToCentimeters(long microseconds)
{
  return microseconds / 29 / 2;
}
```

Let's look at each piece of the code:

❶ Store the LED pins to global variables.

❷ Define a maximum distance of 20cm. If the object moves closer, we will light up an LED next to the corresponding sensor.

❸ The pinMode() function sets a digital pin to act as an output or input. On these lines, we set all digital pins connected to our LEDs to OUTPUT mode.

❹ Within the loop, we call ping() for each sensor and its corresponding LED.

❺ Our new function, ping(), will light up an LED if the measured distance is less than 20cm. With this function we will test the left, middle, and right sensors, one by one.

The function ping() requires two parameters: the sensor (pingPin) and its neighboring LED (ledPin). For example, in the case of the right-side sensor, the pingPin gets a value of 4 and the ledPin gets a value of 7.

❻ Now we measure the distance from the sensor called and store it in variable d. To do so, we pass the pin number associated with a given sensor to the getDistance() function, which returns the measured distance.

❼ Here we declare a Boolean variable called pinActivated and set it to false. Just as the LEDs will show us if a sensor is activated, these Booleans will store true or false statements for each sensor accordingly. In the next section, we will use pinActivated to determine what direction our hand is moving.

❽ If the distance is shorter than maxD, which is set to 20, we will light up the LED (HIGH, +5V) and return a true statement. Otherwise, we will switch off the LED (LOW, 0V), and return false.

WHY TEST WITH LEDs?

The final Interactive Painting will not necessarily include LEDs. So why did we build a temporary version using them? LEDs serve as a practical illustration of how sensors detect motion. By using them, we can find a correct value for distance. Furthermore, detecting the direction of movement, which we'll implement in the final code, is much easier after we can see concretely from the LEDs how the proximity threshold is crossed.

In short, building a practical testing environment speeds up prototyping development. We can more clearly demonstrate the function of a program with LEDs, beeping devices, automatic tests, or by printing information to the serial console.

Determining Direction with the Final Sensor

This last sensor will inform the serial port, using one letter, whether the user has waved her hand from the left to the right (B) or from the right to the left (F).

Building on the earlier examples, this new program detects the direction of the hand wave based on the numbers measured. The challenge is to make the detection so intuitive that any layperson interacting with the painting can learn how to control it within a few tries.

Because the goal is to create an intuitive control method, we need to filter some of the numbers measured to account for factors like unintended hand motions. For example, we'll even out major fluctuations by using an average, removing values below and above the threshold range by setting a waiting period.

This new program uses the three now-familiar LEDs from the previous versions of the code.

```
/* Interactive Painting - Detect direction of hand wave
 Based on code by
 David A. Mellis, Tom Igoe,
 Kimmo Karvinen, and Tero Karvinen
 Updated by Joe Saavedra, 2010
 http://BotBook.com
 */

int slide = 0;

boolean left=false;
boolean center=false;
boolean right=false;

int leftPing = 2;
int centerPing = 3;
int rightPing = 4;

int ledPin = 13;
int leftLedPin = 5;
int centerLedPin = 6;
int rightLedPin = 7;

int maxD = 20; // cm

long int lastTouch = -1; // ms
int resetAfter = 2000; // ms
int afterSlideDelay = 500; //ms; all slides ignored after successful slide
int afterSlideOppositeDelay = 1500;
// left slides ignored after successful right slide

int SLIDELEFT_BEGIN      = -1; // Motion was detected from right
int SLIDELEFT_TO_CENTER  = -2; // Motion was detected from right to center

int SLIDENONE = 0;             // No motion detected

int SLIDERIGHT_BEGIN      = 1;  // Motion was detected from left
int SLIDERIGHT_TO_CENTER = 2;  // Motion was detected from left to center
```

```
void setup() {
  Serial.begin(9600); // bit/s
  pinMode(leftLedPin, OUTPUT);
  pinMode(centerLedPin, OUTPUT);
  pinMode(rightLedPin, OUTPUT);
}

void loop() {
  left = ping(leftPing, leftLedPin); ❶
  center = ping(centerPing, centerLedPin);
  right = ping(rightPing, rightLedPin);

  if (left || center || right) { ❷
    lastTouch=millis(); ❸
  }

  if (millis()-lastTouch>resetAfter ) { ❹
    slide=0; ❺
    digitalWrite(ledPin, LOW);
    // Serial.println("Reset slide and timer. ");
  }

  if (slide >= SLIDENONE) { // only if we are not already in opposite move ❻
    if ( (left) && (!right) ) ❼
      slide = SLIDERIGHT_BEGIN;
    if (center && (slide == SLIDERIGHT_BEGIN)) ❽
      slide = SLIDERIGHT_TO_CENTER;
    if (right && (slide == SLIDERIGHT_TO_CENTER))
      slideNow('R'); ❾
  }

  if (slide <= SLIDENONE) {
    if (right && (!left))
      slide = SLIDELEFT_BEGIN;
    if (center && slide == SLIDELEFT_BEGIN)
      slide = SLIDELEFT_TO_CENTER;
    if (left && slide == SLIDELEFT_TO_CENTER) {
      slideNow('L');
    }
  }
  delay(50);
}

boolean ping(int pingPin, int ledPin) {  ❿

  int d = getDistance(pingPin); //cm  ⓫
  boolean pinActivated = false;
  if (d < maxD) {
    digitalWrite(ledPin, HIGH);
    pinActivated = true;
  }
  else {
    digitalWrite(ledPin, LOW);
    pinActivated = false;
  }
  return pinActivated;
}
```

```
int getDistance(int pingPin) {

  long duration, inches, cm;
  pinMode(pingPin, OUTPUT);
  digitalWrite(pingPin, LOW);
  delayMicroseconds(2);
  digitalWrite(pingPin, HIGH);
  delayMicroseconds(5);
  digitalWrite(pingPin, LOW);
  pinMode(pingPin, INPUT);

  duration = pulseIn(pingPin, HIGH);

  inches = microsecondsToInches(duration);
  cm = microsecondsToCentimeters(duration);

  //return(inches);
  return(cm);
}


void slideNow(char direction) {
  if ('R' == direction)  ⓬
    Serial.println("F");
  if ('L' == direction)
    Serial.println("B");
  digitalWrite(ledPin, HIGH);
  delay(afterSlideDelay);  ⓭
  slide = SLIDENONE;
}


long microsecondsToInches(long microseconds) {

  return microseconds / 74 / 2;
}

long microsecondsToCentimeters(long microseconds) {

  return microseconds / 29 / 2;
}
```

When the user wants to slide one image forward, he waves his hand from right to left, activating the sensors one by one: right, middle, and left. The program sends a slide command F to the serial port when all the sensors are triggered.

Let's examine the code:

❶ Here we measure whether the left sensor has been triggered, and then measure for the center and right sensors. The ping() function returns a true/false value (of the boolean type) indicating whether the sensor has been triggered (that is, whether the object is closer than the threshold value of 20cm). A true value means that the sensor has been triggered, and false means the opposite. The program then checks in the same way whether the middle or right sensors have been triggered.

❷ Because the value returned by `ping()` is a `true`/`false` (Boolean) value, it can be compared using Boolean operators, such as *OR* (||) and *AND* (&&). The comparison here means, "If the left sensor has been triggered OR the middle sensor has been triggered OR the right sensor has been triggered, THEN…." For example, if only the left sensor has been triggered (has a `true` value), the comparison will be equivalent to this:

```
if (true || false || false)
```

❸ If even one sensor has been triggered, store the current time in milliseconds in the `lastTouch` variable. Later in the program, we can compare the value returned by `millis()` to the value of `lastTouch` to determine the amount of time that has passed since this point.

❹ From this moment of initialization, calculate the time since the `lastTouch` moment. The time elapsed is the current `millis()` moment minus the earlier `lastTouch` moment.

❺ If the time period since `lastTouch` is longer than the `resetAfter` value, set `slide` to 0. The beginning of the code requires the triggers within one slide (wave of a hand) to take place within two seconds (`resetAfter` is 2000).

❻ First, check that the user's hand is not moving in the other direction (`slide>=0`)—in other words, `slide` is either 0 or positive (1 or 2). In a neutral situation, `slide` is 0, which means that either the program was just started or the counter (explained in Step 12) was just zeroed.

❼ If the left sensor has been triggered but the right one hasn't, move one step toward the slide.

❽ If we have previously moved one step closer to the slide, and the middle sensor has been triggered, we will move one more step toward the slide.

❾ If we are only one more step from the slide, and the right sensor is triggered, we will execute the slide by calling `slideNow()`.

❿ The `ping()` function indicates whether the waving hand is within a certain distance of the sensor. To filter out glitches, this function takes two measurements and uses their average.

⓫ First we take a measurement from the sensor, and then compare it to our `maxD` (maximum distance) variable. This serves as the threshold of distance for triggering the sensor.

⓬ If the motion is forward (`'R'` for right), `slideNow()` prints the character F to the serial port (`'L'` stands for left, or backward motion, so a B is printed), and resets the counter to 0 to wait for a new motion.

⓭ The program waits for a half a second (the duration defined in the beginning of the code) without reacting to new events. This way, minor hand twitches do not move the images back and forth.

Sliding images backward works the same way as sliding images forward.

Here, after the `if` statement, we do not have curly braces ({}) symbolizing a block, since the block consists of only one line: `slide=1;`. If you add another line to this `if` statement, you should also add the curly braces. Then the next line will also belong to the block after the `if` statement.

The variable `slide` serves as a counter that increases when the sensors are triggered from right to left, and decreases in the opposite direction. For the `slideNow()` function to be called, the variable `slide` will have reached either 2 or –2, depending on the direction. Here, using 0 as our neutral point, we can determine which direction to send the images. For example, when `slide`'s value is 2, `slideNow()` will send an F, triggering the slideshow to move forward, and then zero the counter.

When we were developing this program, we had to come up with values for variables such as `afterSlideDelay`. To do this, we tested the device with LEDs attached to it, recruited people nearby as testers, and followed the movement. Then we simply improved the motion detection and tried with the next volunteer.

Moving Images

Now we'll create a program that moves the images in the necessary direction (Figure 5-14). We'll do this first with Python and then with Processing. You can decide which language better fits your needs.

Figure 5-14. *Using a gesture to move images*

Installing Python

Before going any further, you'll need to install Python. The following sections detail how to get up and running in Linux, Mac OS X, and Windows.

Linux and Mac OS X

Python and PyGTK are already installed under most Linux distributions, so you can just start programming. If they're"n"t already installed, you can install them using your Linux distribution's package manager. On Mac OS X, you can use MacPorts (*http://www.macports.org/*). After you install MacPorts, run the command sudo port install py-gtk2 at the Terminal (*/Applications/Utilities/Terminal*) to install PyGTK.

Older Ubuntu Linux installations include a demonstration of PyGTK's capabilities (Figure 5-15). Ubuntu 9.04 enables you to install the demo separately yourself. Open a Terminal window via Applications→Accessories→Terminal and issue the following commands:

```
$ sudo apt-get update
$ sudo apt-get install python-gtk2-doc
```

You can run the demo on Mac or Linux by running this command in a Terminal window:

```
$ pygtk-demo
```

Figure 5-15. *PyGTK demo in Ubuntu Linux*

Windows 7

One way to install PyGTK on Windows is to use Cygwin (*http://www.cygwin.com*), an open source system that brings many Unix and Linux applications to Windows, including such things as the Bash shell and the X Window System. However, Cygwin installs a lot of software you don't need for this project, so we'll use the native Windows version of PyGTK.

Download the installation packages

You'll need to download Python 2.6, as well as several other installers. You can find all files from *http://BotBook.com*. There are newer versions available, but this project was tested with Python 2.6 and the following packages (shown in Figure 5-16):

Python for Windows
    http://www.python.org/download/releases/2.6.5/

The GTK+ Library (download the gtk+-bundle file)
    http://ftp.se.debian.org/pub/gnome/binaries/win32/gtk+/2.14/

PyCairo (download the most recent .exe file that has py2.6 in its name)
    http://ftp.gnome.org/pub/GNOME/binaries/win32/pycairo/1.4/

PyObject (download the most recent .exe file that has py2.6 in its name)
    http://ftp.gnome.org/pub/GNOME/binaries/win32/pygobject/2.14/

PyGTK (download the most recent .exe file that has py2.6 in its name)
    http://ftp.gnome.org/pub/GNOME/binaries/win32/pygtk/2.12/

Log in as an admin user

If you haven't already logged in as a user with administrator privileges, you should do so now. That way, you'll be able to install all these files without any difficulty.

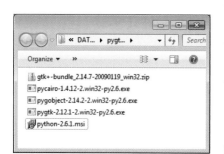

Figure 5-16. *The installation files downloaded to your computer*

Install Python

Run the *python-2.6.x.msi* installer, making the following choices:

1. Select "Install for all users," and then click Next (Figure 5-17).

2. When prompted to "Select Destination Directory," click Next.

3. When prompted to "Customize Python 2.6.x," click Next.

4. Click Finish.

Install PyGTK and other libraries

Run the *pycairo-1.4.12-x.win32-py2.6.exe* installer and do the following:

1. When the installer tells you "This Wizard will install pycairo," click Next.

2. When you see the message "Python 2.6 is required," make sure it finds the Python installation you installed in the previous step. Click Next.

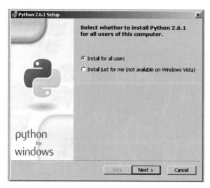

Figure 5-17. *Click Next to proceed through the installation*

3. When prompted to "Click Next to begin the installation," click Next (Figure 5-18).

4. Click Finish.

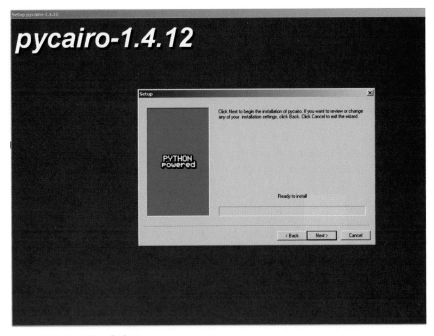

Figure 5-18. *Installing PyCairo*

Run the *pygobject-2.14.2-x.win32py2.6.exe* installer and do the following:

1. When the installer tells you "This Wizard will install pygobject," click Next.

2. When you see the "e"sage "Python 2.6 is required," make sure it finds the Python installation you installed earlier. Click Next.

3. When prompted to "Click Next to begin the installation," click Next.

4. Click Finish.

Run the *pygtk-2.12.1-x.win32-py2.6.exe* installer and do the following:

1. When the installer tells you "This Wizard will install pygtk," click Next.

2. When you see the message "Python 2.6 is required," make sure it finds the Python installation you installed earlier. Click Next.

3. When prompted to "Click Next to begin the installation," click Next.

4. Click Finish.

Install the GTK+ Library

Double-click the *gtk+-bundle_2.14.x-x_win32.zip* file that you downloaded earlier and then complete these steps:

1. Create the following folder: *C:\Program Files\gtkbundle* (Figure 5-19).

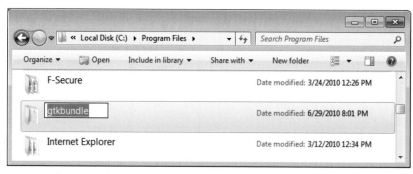

Figure 5-19. *Creating a folder for PyGTK*

2. When you double-click the zip file, a new Explorer window opens. Select all the folders in that window (*bin*, *contrib*, *lib*, and so forth) and drag (copy) them into *C:\Program Files\gtkbundle*.

3. Add the *bin* directory to your computer's Path variable as follows (Figure 5-20):

    a. Open the Start menu, locate your Computer, and right-click. Choose Properties from the menu that appears. Locate the Advanced System Settings icon and click it.

    b. Click the button labeled Environment Variables. Locate the Path entry under the "System variables" section and click Edit.

    c. Append the following (do not change anything else in the Path variable and do not add any extra space; if you make a mistake, click Cancel and start over):

    ```
    ;C:\Program Files\gtkbundle\bin;C:\python26
    ```

    d. Click OK to dismiss the remaining windows.

    e. Close any previously opened command prompts (the next time you start the command prompt, it will have the new Path setting).

4. Test the installation. Open a command prompt (Start→All Programs→Accessories→Command Prompt). Type **gtk-demo** and press Enter (Figure 5-21).

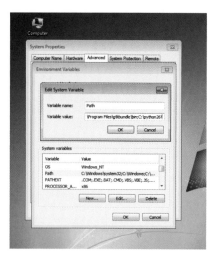

Figure 5-20. *Adding PyGTK's bin directory to the Path variable*

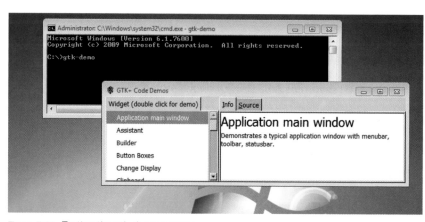

Figure 5-21. *Testing the gtk-demo program*

The gtk-demo program opens a window where you can explore GTK+'s features. If you aren't able to launch it, double-check the previous instructions to make sure you've installed everything correctly. If you still have trouble, visit *http://www.gtk.org/download-windows.html* for more information about running GTK+ on Windows.

If you had to log in as an administrator to install these files, you can now log out and log back in as a standard user.

PyGTK is now installed and you can begin to write programs with it.

Hello World in Python

Hello World is the simplest, first, and most important program we'll use to begin all of our projects.

Open a text editor. Within Ubuntu Linux, you can find Gedit in Applications→Accessories→Text Editor. On Windows, you can use Notepad (Figure 5-22): Start→Programs→Notepad. Mac OS X includes TextEdit (*/Applications/TextEdit*). Alternatively, you might want to try an open source programmer's text editor such as jEdit (*http://www.jedit.org/*) or Geany (*http://geany.org*), which runs on Mac, Windows, and Linux.

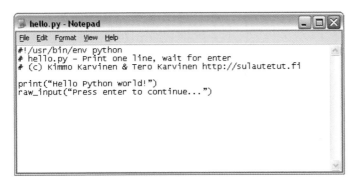

Figure 5-22. Writing code in Notepad

Copy the following code to your text editor (or download it from the book's website, *http://BotBook.com/*) and save it with the name *hello.py*:

```
#!/usr/bin/env python ❶
# hello.py - Print one line, wait for enter
# (c) Kimmo Karvinen & Tero Karvinen http://BotBook.com
print("Hello Python world!") ❷
raw_input("Press enter to continue...") ❸
```

Execute the program. If you're using Windows, double-click the hello.py icon, and the program will print a familiar welcome message. Alternatively, you can run the program from the command prompt by going to Start→All Programs→Accessories→Command Prompt. Navigate to the folder where you saved the file (using the cd command) and execute the program with the command python hello.py.

If you're running Linux, open the Terminal (Applications→Accessories→ Terminal) and navigate to the folder (using the cd command) where you saved *hello.py*.

It is not easy to save simple text files (files without formatting) with office programs, so don't try to edit the code with OpenOffice or Microsoft Word.

Change the privileges (chmod) by adding (+) the execution rights (x) for the *hello.py* file:

```
$ chmod +x hello.py
```

Then, run the program by double-clicking its icon or running the following command:

```
$ ./hello.py
```

When you execute the program, it will print the following text to your screen:

```
Hello Python world!
Press enter to continue...
```

You have just written the most important beginning Python program!

Let's review the Hello World program in the order of execution:

❶ Tell Unix-based systems (Mac OS X and Linux) where to find the Python program itself. If the program were missing this line, you could still run it under Windows, Linux, and Mac OS X by executing it from the command line with the command python hello.py.

After the shebang sign (#!), the name of the command interpreter (python) appears. When you see this line on Linux or Mac systems, there is no need to type the interpreter python hello.py in front of the program name; just typing ./hello.py is sufficient.

❷ Print the text "Hello Python world!" to the screen.

❸ Print the text "Press enter to continue…" and wait for the user to press Enter. Adding this line allows Windows users to see the program output when they double-click the icon to launch it (otherwise, it would pop up on the screen and disappear quickly).

Hello Windows

Have you always wanted to make windows and buttons? Wouldn't it be nice if the same code to create them worked in Linux, Windows, and Mac OS X?

Let's make a window within Python. Open a text editor, write the following code, and save it with the name *windowHello.py*:

```
#!/usr/bin/env python
# windowHello.py - Create a window with a button.
# (c) Kimmo Karvinen & Tero Karvinen http://BotBook.com
import gtk ❶

window = gtk.Window() ❷
window.connect("destroy", gtk.main_quit) ❸

button = gtk.Button("Hello PyGTK - BotBook.com") ❹
window.add(button) ❺

window.show_all() ❻

gtk.main()
```

Within Windows, double-click the helloWindow.py icon to execute the program. On Linux or Mac OS X, make the program into a command-line executable:

```
$ chmod +x helloWindow.py
```

If you used MacPorts to install PyGTK, you need to modify all the Python scripts in this chapter to refer to the appropriate MacPorts version of Python in the #! line, changing:

```
#!/usr/bin/env python
```

to one of the following, depending on which version of Python MacPorts installed during the PyGTK installation:

```
#!/opt/local/bin/python2.4
#!/opt/local/bin/python2.5
#!/opt/local/bin/python2.6
```

Then, execute it by double-clicking the program icon or via the command line (the $ is the shell prompt; type everything that comes after it):

```
$ ./helloWindow.py
```

A small window with a button will open on the screen with the message "Hello PyGTK - BotBook.com," as shown in Figure 5-23. The button will not do anything, but you can quit the program by closing this window.

Let's examine the components of the Hello Windows script:

Figure 5-23. *Hello Python window*

❶ Load the PyGTK program library for displaying windows and user interfaces. This library is known as *gtk*, *python.gtk*, and PyGTK.

❷ Create a new window. Save it into a variable called `window`.

❸ Send a `destroy` event and call a `gtk.main_quit()` function when the user clicks the window's close (X) button. The function is passed the event and the button as its parameters.

❹ Create a new button and add it to the window.

❺ Show the window and its contents.

❻ Start the GTK's main loop, where the program will spend the rest of its execution time while waiting for events.

This new program defines only one event: the window close (X) button that ends the main loop and, therefore, the whole program.

Communicating over the Serial Port

Open the Arduino development environment and use the following test program to make sure the development environment's serial console is working properly:

```
// printSerial.pde - Print test data to serial port.
// (c) Kimmo Karvinen & Tero Karvinen http://BotBook.com

void setup()
```

```
{
    Serial.begin(9600); // bit/s ❶
}

void loop()
{
    Serial.print("F "); ❷
    delay(500);
    Serial.print("B ");
    delay(1000);
    Serial.print(" FFFF ");
    delay(500);
    Serial.print(" http://botbook.com ");
    delay(2000);
}
```

Load this program onto your Arduino and then launch the Serial Monitor (Tools→Serial Monitor). The program will repeatedly print the text "FFFF botbook.com F B." You can later use the same program to simulate recognized hand movements.

❶ The serial port must be opened before writing. The program opens it at 9,600 bits per second.

❷ Write a string of text to the serial port.

Installing the PySerial Library

The PySerial library is required for interacting with the serial port from Python. After you install it, you'll be able to create Python scripts that can talk to Arduino.

Linux and Mac OS X

In Ubuntu Linux, it's easiest to install the PySerial library via the command line:

```
$ sudo apt-get install python-serial
```

> *If you can't find this package on Ubuntu, you'll need to enable the third-party open source repository named Universe and update the available software package list by running these commands at the Terminal:*
>
> ```
> $ sudo software-properties-gtk --enable-component=universe
> $ sudo apt-get update
> ```
>
> *Other Linux distributions might have different procedures; if all else fails, you can install it from source (http://pyserial.sourceforge.net/).*

On Mac OS X, if you are using MacPorts (see the note about this in the "Linux and Mac OS X" section in "Installing Python"), use the command:

```
$ sudo port install py-serial
```

Windows XP

In Windows, the PySerial library also requires the PyWin32 library as its support. In addition to the home pages mentioned here, you can download the programs at *http://BotBook.com*.

If necessary, log in as a user with administrative privileges.

Go to the PyWin32 program home page at *http://sourceforge.net/projects/ pywin32/* and download the latest version of the Python Extensions for Python 2.6. The file will be named something like *pywin32-version.win32-py2.6.exe*. Install the file by double-clicking its icon.

Next, download the PySerial library. Go to *http://sourceforge.net/projects/ pyserial/files/* and download the latest version for Windows. Install it by double-clicking its icon.

If you had to log in as an administrator to do this, you can log out now and log back in as a standard user.

> *Unless you have Python installed in a strange place, you can accept the defaults for each step of these installations.*

Reading the Serial Port

Let's write a Python program that reads the serial port and prints to the screen.

Open a text editor (such as Notepad, gEdit, TextEdit, or another plain-text editor). Copy the following program code and save it under the filename *read Serial.py*:

```
#!/usr/bin/env python
# readSerial.py - Read serial port and print to screen
# (c) Kimmo Karvinen & Tero Karvinen http://BotBook.com

import serial, sys ❶

# File name will be different if you are on Windows or Mac OS X
ser = serial.Serial("/dev/ttyUSB1", 9600) ❷

if (ser):
    print("Serial port " + ser.portstr + " opened.")

while True: ❸
    sys.stdout.write ❹(ser.read(1) ❺)
    sys.stdout.flush() ❻
```

Let's review the code one segment at a time:

> *You must change /dev/ttyUSB1 to the filename of the serial port your Arduino is connected to, which you can find from the development environment menu (Tools→Serial Port).*
>
> *Only one program can use the serial port at a time, so don't try to use the Arduino Serial Monitor while you are running the Python program. Within Windows, serial ports are named in the format COMX, where X is replaced by some number. Within Linux, serial ports are device files such as /dev/ttySX or dev/ttyUSBX. On Mac OS X, they are named /dev/ tty.usbserial-XXXXX.*

❶ Import the necessary libraries. Naturally, the most important library is PySerial, the serial library we are using for the serial port. The library called *sys* allows us to print text to the screen without automatically adding a carriage return.

❷ Open the serial port. This creates a new `Serial` class object called `ser`. You must change */dev/ttyUSB1* to the name of your Arduino serial port. You can find out which serial port you are using for Arduino by going to Tools→Serial Port.

❸ This next block is repeated forever. End the program by killing it, closing its window, or pressing Ctrl-C.

❹ Write one character without a carriage return. Because we don't want any extra carriage returns, we won't use the usual `print()` command.

❺ Read 1 byte from the serial port. One byte corresponds to one character— for example, A, *, or 2.

⑥ On systems that use buffered output, make sure the text is displayed immediately.

Here is some sample output from the program:

```
Serial port /dev/ttyUSB1 opened.
  B
  FFFF
http://botbook.com
  F
```

What do you do if the program doesn't function? Make sure that no other program has taken over the serial port, including any of the other programs used in this project. The Arduino development environment serial console and serproxy must also be closed. Finally, make sure you are reading the correct serial port. Can you read the same serial port in the Arduino development environment serial console?

Reading the Serial Port and PyGTK

Although this program uses PyGTK, for the sake of simplicity, it will display its output in text mode and will not create windows. (When testing features in stages, it is best to test only one feature at a time.)

If you installed Python under Mac OS X using MacPorts, be sure to review the earlier note in the "Linux and Mac OS X" section of "Installing Python."

```
#!/usr/bin/env python
# gtkserial.py - Read serial port with GTK.
# (c) Kimmo Karvinen & Tero Karvinen http://BotBook.com
import serial, gtk, gobject, sys ❶

# File name will be different if you are on Windows or Mac OS X
ser = serial.Serial('/dev/ttyUSB1', 9600) ❷

def pollSerial(): ❸
    sys.stdout.write(ser.read(1))
    sys.stdout.flush()
    return True

if (ser):
    print("Serial port " + ser.portstr + " opened.")
    gobject.timeout_add(100, pollSerial) ❹

gtk.main()
```

Be sure to change /dev/ttyUSB1 to the filename of the serial port your Arduino is connected to.

This program outputs text just like the previous console-based program.

Let's view our program one segment at a time. Most of the program is already familiar to you from the previous serial port example, so we'll concentrate on the differences.

❶ Load the graphical user interface libraries *gtk* and *gobject*.

❷ Change */dev/ttyUSB1* to the name of your Arduino serial port. You can find out which serial port you are using for Arduino by going to Tools→Serial Port.

❸ The pollSerial() function returns a true value, and the timer function (described next) calls pollSerial() again and again. The function's only command reads 1 byte from the serial port and outputs it to the screen.

❹ The timer function `timeout_add()` takes time in microseconds and a called function as its parameters. This timer calls the `pollSerial()` function after 0,1 seconds, as long as it returns a `true` value.

Displaying a Picture

The central task of this program is to load a picture from a file and display it in a window. The following code uses an example picture (*data/image1.jpg*), but you can use any JPG image for the file; just replace *image1.jpg* with your filename and *data* with the directory that contains it.

```
# imageHello.py - Display image in a window.
# (c) Kimmo Karvinen & Tero Karvinen http://BotBook.com

import gtk, os ❶

window = gtk.Window() ❷
window.connect("destroy", gtk.main_quit)

image=gtk.Image() ❸
window.add(image)
image.set_from_file(os.path.join("data", "image1.jpg")) ❹

window.show_all() ❺
gtk.main()
```

Let's review the code in segments:

❶ Import the graphical user interface library *gtk* (PyGTK) and the *os* library, which can understand the intricacies of different operating systems (see the upcoming Note for an example).

❷ These two lines create a normal window and enable its window close button (usually an X in one corner of the window). We will refer to the window with a variable named `window`, but the window won't appear yet.

❸ These two lines create a new empty image object called `image`. The picture is added to the window, making it the only widget there.

❹ Load a picture onto the `image` object from the *image1.jpg* file in the *data* folder.

❺ Display the window object named `window` with all its widgets. From here on out, the program spends all of its time in the main `gtk` loop, where `gtk` waits for events. When the user triggers the only defined event (clicking the window close button), the program ends.

Different operating systems have different directory separators. Windows uses a backslash: \. Linux, Unix, and Mac OS X use a forward slash: /. Luckily, you don't have to worry about these differences with Python. The `os.path.join()` function combines different sections of the directory with the correct characters. After the `os.path.join()` function has combined the directories, the path will be data/image1.jpg under Linux and Mac OS X, and data\image1.jpg under Windows.

Scaling an Image to Full Screen

It seems natural to show the picture in full screen in our final program. The following program scales the image to be as large as possible while maintaining the aspect ratio. It simultaneously makes a copy of the picture, because we'll need this type of picture later in the project for animating the slides.

Save the following file as *fullScreenScale.py*:

```python
#!/usr/bin/env python
# fullScreenScale.py - Show image scaled to full screen
# (c) Kimmo Karvinen & Tero Karvinen http://BotBook.com

import gtk, os

def fitRect(thing, box): ❶
    # scale ❷
    scaleX=float(box.width)/thing.width
    scaleY=float(box.height)/thing.height
    scale=min(scaleY, scaleX) ❸
    thing.width=scale*thing.width ❹
    thing.height=scale*thing.height
    # center ❺
    thing.x=box.width/2-thing.width/2
    thing.y=box.height/2-thing.height/2 ❻
    return thing ❼

def scaleToBg(pix, bg): ❽
    fit=fitRect( ❾
        gtk.gdk.Rectangle(0,0, pix.get_width(), pix.get_height()),
        gtk.gdk.Rectangle(0,0, bg.get_width(), bg.get_height())
    )
    scaled=pix.scale_simple(fit.width, fit.height, gtk.gdk.INTERP_BILINEAR) ❿
    ret=bg.copy() ⓫
    scaled.copy_area( ⓬
        src_x=0, src_y=0,
        width=fit.width, height=fit.height,
        dest_pixbuf=ret,
        dest_x=fit.x, dest_y=fit.y
    )
    return ret ⓭

def newPix(width, height, color=0x000000ff): ⓮
    pix=gtk.gdk.Pixbuf(gtk.gdk.COLORSPACE_RGB, True, 8, width , height) ⓯
    pix.fill(color) ⓰
    return pix ⓱

def main(): ⓲
    pix=gtk.gdk.pixbuf_new_from_file(os.path.join("data", "image1.jpg")) ⓳
    window = gtk.Window()
    window.connect("destroy", gtk.main_quit)
    window.fullscreen() ⓴

    bg=newPix(gtk.gdk.screen_width(), gtk.gdk.screen_height()) ㉑
    pixFitted=scaleToBg(pix, bg) ㉒

    image=gtk.image_new_from_pixbuf(pixFitted) ㉓
    window.add(image)
    window.show_all()
    gtk.main()

if __name__ == "__main__": ㉔
    main()
```

The program is similar in many ways to the previous image-presenting programs, so let's look only at the differences. The program execution begins with main(), so you might want to skip down there first:

❶ The function receives rectangles, as defined in the call, as its parameters. They are therefore of the type `gtk.gdk.Rectangle`. For example, if the call includes image dimensions of 125×200, `thing.height==200` and `thing.width==125`. The rectangle `thing` also has the upper-left corner coordinates `thing.x==0` and `thing.y==0`.

❷ Calculate the image height as a percentage of the image width. For example, a background with a width of 500 pixels is 500% of the width of an object 100 pixels wide. Therefore, `scaleX` receives a value `5,0`. We compare the height the same way. In division, the numerator (upper number) must always be converted into a `float`. When dividing with integers, you can get totally preposterous end results. For example, `1/2` will erroneously return the end result of `0`, but the float `(1)/2` will correctly return `0,5`.

❸ Choose a smaller scaling for the scaling of the whole image. Height and width must be scaled with the same number to preserve the image aspect ratio. Extra space will be filled with black.

❹ Scale the object height and width, storing the new dimensions over the old ones. Now the `thing` variable has the correct dimensions.

❺ We'll place the object in the center. We first need to calculate how far from the upper-left corner (0,0) of the background (`bg`) to place the object. To center the object, we need to place its midpoint at the midpoint of the background (in other words, at half the width of the background: `box.width/2`).

The distance between the center point of the object and the left side of the object is `thing.width/2`. Therefore, the upper-left corner of the object is set at half of the width of the object subtracted from half of the width of the background.

❻ Calculate the position of the top of the image in the same way as the side of the image.

❼ Finally, return the location and size of the object. The variable to be returned is of the type `gtk.gdk.Rectangle`—for example, `thing.x==123`, `thing.y==22`, `thing.width=1024`, `thing.height=768`. When attached to the background, the object that just fits in the frame should be located at the coordinates (123, 22) and scaled to size 1024×768. The coordinates are calculated from the upper-left corner of the background.

❽ Pass two Pixbuf image buffers to the function as its parameters. Pixbuf is an image stored in the computer memory that can be effectively modified and stretched. Though we use the same names as in the main program for the parameters of the function, they are naturally different variables. Within the `scaleToBg()` function, `pix` and `bg` can be any image buffers defined in the calling of the function.

❾ These lines calculate how to fit the image within `bg` without altering the dimensions of the image and call the function `fitRect()` to accomplish this. We define the image dimensions by using the GTK library `Rectangle` object.

The `Rectangle` object includes the coordinates for the upper-left corner (x, y) and the width and the height. If, for example, the size of the image `pix` is 125×200, the first square is equivalent to `gtk.gdk.Rectangle (0,0, 125, 200)`.

We don't use the upper left-corner coordinates for anything here, so we leave them set as zeros.

❿ Stretch `pix` to the right size and save the result into a new pixel buffer named `scaled`. The parameters of the method `scale_simple` are the new width and height and the scaling method. Scaling methods vary in speed and quality. Bilinear scaling is a good compromise.

⓫ Create a copy of the background `bg` to protect the original one. We store the copy in a new variable called `ret`.

⓬ Copy the previously scaled Pixbuf over the scaled picture `ret`. The source image for the copying here is `scaled`, with `copy_area()` as its method. The `copy_area()` method will likely have quite a number of parameters, so in the call we use an option provided by Python to print the parameter names—for example, `dest_pixbuf` or `src_x`. Most other languages, such as C, do not have such an option.

The area copied from the source image (`scaled`) is defined as coordinates (`src_x`, `src_y`) as well as width and height. Coordinates are defined from the source image (`det_pixbuf`), onto which a copy of the source image is placed at (`dest_x`, `dest_y`).

We defined earlier the dimensions and upcoming position of the image to be copied with the `fitRect()` function. Here, `fit.width` and `fit. height` are the same as the dimensions of the scaled image.

⓭ Finally, `scaleToBg()` returns an image scaled to its background. Since the background `bg` is the same size of the screen, the returned Pixbuf is also the size of the screen. Once it executes, the program returns to the `main()` function.

⓮ The `newPix()` function takes an image width and height as its parameters. If needed, the caller can set the image color with a third parameter; black is the default.

⓯ Create a new image buffer, `Pixbuf`, and store it in the variable `pix`. Within the call of a `Pixbuf` class constructor, we define a color palette, transparency, color depth, and dimensions. The color palette used is `gtk.gdk.COLORSPACE_RGB` and the color depth is 8. Transparency (alpha channel) is also available, and is discussed next.

⓰ Fill the image buffer with an even color. If the user has not defined color as a parameter, use black (`0x000000ff`) as the default value. The color definition shown here is in hexadecimal, a base-16 number system. Colors are given in RGB (red, green, blue) format. When all of the values are zeros, the color is black. When all are `0xff` (16x16=256), the color is white. That gives us six characters (two each for R, G, and B), but there are eight shown here. The last two characters are used for defining the opacity of a color. The largest number, `0xff` (256), means that the color is fully opaque. Similarly, 00 means that the color is fully transparent (invisible).

⑰ The function returns a new image buffer, and execution of the program returns to main(), storing the black image buffer we created here in a variable bg.

⑱ We wrote the main program as its own function. There are many benefits to using this structure in a long program like this one. For example, not all the variables of the main program need to be global variables, so putting the program in its own function isolates those variables.

⑲ Read the image from the disk. We will connect the folder and filenames with the os.path.join() function, so we won't have to worry about operating system–specific differences. This line also loads the image to a new image buffer (Pixbuf). We can use Pixbuf only by copying it to a gtk.Image type image widget. We will store the image buffer in a new object named pix.

⑳ It's easy to display the image in full screen. At the same time, it is useful to offer a simple way for the user to close the program without a window-closing X button. This example program does not do so. The user must close the program with Alt-F4 (close a window) or switch to another program by pressing Alt-Tab or Command-Tab.

㉑ Create a new black background image and store it in variable bg. The image is the same size as the screen. We check the dimensions with the screen method. Our window has not appeared yet, so we can't easily check its size. If, for example, the display is 1024×768, the function call will be equivalent to bg=newPix(1024, 768).

㉒ Stretch the image to fit the background using the scaleToBg() function, which makes the image as large as possible and leaves a border around it. The image buffer it returns is therefore exactly as high and as wide as bg.

㉓ The next three lines create an image object from the pixels in the pixFitted object, add the image to the window, and display the scaled image by showing the window.

㉔ This line is a Python trick that invokes the main() method only if the file was run directly (such as from the command line). It enables us to turn this program into a library and use its functions by importing the program as a library: import fullScreenScale.

You know now how to scale images. Though you need to be careful with calculations, you don't have to think about them too often, and you can use the same image-stretching function in all programs.

Changing Images with Button Control

Here's a program that displays all images from a folder. The user can change images from the keyboard using the space bar and the B key.

```
#!/usr/bin/env python
# multipleImages.py - Display image in a window.
# (c) Kimmo Karvinen & Tero Karvinen http://BotBook.com

import gtk, os

dir="data"  ❶
pixbufs=[]
image=None
pos=0

def loadImages():
    for file in os.listdir(dir):  ❷
        filePath=os.path.join(dir, file)  ❸
        pix=gtk.gdk.pixbuf_new_from_file(filePath)  ❹
        pixbufs.append(pix)  ❺
        print("Loaded image "+filePath)

def keyEvent(widget, event):  ❻
    global pos
    key = gtk.gdk.keyval_name(event.keyval)  ❼
    if key=="space" or key=="Page_Down":  ❽
        pos+=1
        image.set_from_pixbuf(pixbufs[pos])
    elif key=="b" or key=="Page_Up":
        pos-=1
        image.set_from_pixbuf(pixbufs[pos])
    else:
        print("Key "+key+" was pressed")  ❾

def main():
    global image
    window = gtk.Window()
    window.connect("destroy", gtk.main_quit)  ❿
    window.connect("key-press-event", keyEvent)  ⓫
    image=gtk.Image()
    window.add(image)  ⓬
    loadImages()
    image.set_from_pixbuf(pixbufs[pos])

    window.show_all()
    gtk.main()  ⓭

if __name__=="__main__":
    main()
```

This program is similar to the single-image viewer, so we'll concentrate on the differences. Let's look at each piece of the code.

❶ Declare global variables in the beginning of the program, outside of all functions. Images are read from the *dir* directory and will later be added to the pixbufs list. The pos variable shows the order of the picture shown. We'll fill in the image shown in a window (image) later.

❷ List the contents of our image directory, which we defined previously as *data*. With this variable value, the command will be equivalent to os.listdir("data"), which will return a list, such as ['foo.jpg', 'image1.jpg', 'sulautetut.svg']. This list will be used in the for loop, which would be equivalent to something like for file in ['foo.jpg', 'image1.jpg', 'sulautetut.svg'].

You can download these images from http://BotBook.com. *You might need to install the* librsvg *library to use this example with SVN images. On Mac OS X with MacPorts, use* sudo port install librsvg. *On Ubuntu Linux, use* sudo apt-get install python-rsvg. *On Windows, it may be a bit more complicated (see* http://librsvg.source-forge.net/ *for more information about this library).*

❸ Combine each file with its path, using the `os.path.join()` function.

❹ Load each file into a new pixel buffer (of type `Pixbuf`) called `pix`. The `pixbuf_new_from_file()` makes our life easier by creating a new `Pixbuf` object and loading an image from a file into it at the same time. `Pixbuf` objects are meant for handling images within the computer memory. They cannot be displayed directly. We'll have to copy data from the `Pixbuf` object into an image widget to display it.

❺ Append our image to the end of the list `pixbufs`. The length of the list grows by one. Now the last member of the list is the same as `pix`. When the loop has been executed with all the values, the program returns to the point in `main()` at which `loadImages` was called.

❻ React to two types of events: clicking the window close (X) button and pressing the keys. Pressing a key calls the `keyEvent()` function from the GTK main loop. This function is passed in the widget that caused the event as well as in the event itself.

❼ Read the name of the key press from the event that was passed in.

❽ If the key pressed is a space bar or Page Down, move one image forward and show the corresponding `Pixbuf` object in the widget image. The first position in the list is numbered 0. For example, if we are in the second image, the value will be equivalent to `image.set_from_pixbuf(pixbufs[1])`, which is the picture loaded from the directory *image1.jpg*. The landscape will appear immediately within the image widget, which fills the window. Moving backward works the same way.

❾ Display other keys in the console for informational purposes.

❿ If the user closes the window, the program quits.

⓫ Detect any key presses within the window when the window is active. When a key press event occurs, call the `keyEvent()` function. GTK automatically sends the widget that received the event, as well as the event itself, as parameters.

⓬ Add the first picture in the image list to the image widget shown in the window. Here, the value of variable `pos` is 0 (the first image in the list).

⓭ Start the main loop of the GTK library, where the program will spend the rest of its execution time while waiting for events.

Gesture-Controlled Painting in Full Screen

Now's the time to combine all the components we have built into a painting a user can control by waving her hand in front of it.

```
#!/usr/bin/env python
# handWaveFull.py - Choose full screen image by waving hand.
# (c) Kimmo Karvinen & Tero Karvinen http://BotBook.com

import gtk, os, serial, gobject
```

```
# Global variables

dir="data"
pixbufs=[]
image=None
bg=None
pos=0
ser=None

# Pixbuf manipulation

def fitRect(thing, box):
    # scale
    scaleX=float(box.width)/thing.width
    scaleY=float(box.height)/thing.height
    scale=min(scaleY, scaleX)
    thing.width=scale*thing.width
    thing.height=scale*thing.height
    # center
    thing.x=box.width/2-thing.width/2
    thing.y=box.height/2-thing.height/2
    return thing

def scaleToBg(pix, bg):
    fit=fitRect(
        gtk.gdk.Rectangle(0,0, pix.get_width(), pix.get_height()),
        gtk.gdk.Rectangle(0,0, bg.get_width(), bg.get_height())
    )
    scaled=pix.scale_simple(fit.width, fit.height, gtk.gdk.INTERP_BILINEAR)
    ret=bg.copy()
    scaled.copy_area(
        src_x=0, src_y=0,
        width=fit.width, height=fit.height,
        dest_pixbuf=ret,
        dest_x=fit.x, dest_y=fit.y
    )
    return ret

def newPix(width, height, color=0x000000ff):
    pix=gtk.gdk.Pixbuf(gtk.gdk.COLORSPACE_RGB, True, 8, width , height)
    pix.fill(color)
    return pix

# File reading

def loadImages():
    global pixbufs
    for file in os.listdir(dir):
        filePath=os.path.join(dir, file)
        pix=gtk.gdk.pixbuf_new_from_file(filePath)
        pix=scaleToBg(pix, bg)
        pixbufs.append(pix)
        print("Loaded image "+filePath)

# Controls

def go(relativePos):  ❶
    global pos
    pos+=relativePos
```

```
        last=len(pixbufs)-1 ❷
        if pos<0:
            pos=last
        elif pos>last:
            pos=0

        image.set_from_pixbuf(pixbufs[pos])

    def keyEvent(widget, event):
        global pos, image
        key = gtk.gdk.keyval_name(event.keyval)
        if key=="space" or key=="Page_Down":
            go(1) ❸
        elif key=="b" or key=="Page_Up":
            go(-1)
        elif key=="q" or key=="F5":
            gtk.main_quit()
        else:
            print("Key "+key+" was pressed")

    def pollSerial():
        cmd=ser.read(size=1)
        print("Serial port read: \"%s\"" % cmd)
        if cmd=="F":
            go(1)
        elif cmd=="B":
            go(-1)
        return True

    # Main

    def main():
        global bg, image, ser
        bg=newPix(gtk.gdk.screen_width(), gtk.gdk.screen_height())
        loadImages()
        image=gtk.image_new_from_pixbuf(pixbufs[pos])

        ser = serial.Serial('/dev/ttyUSB1', 9600, timeout=0 ❹)
        gobject.timeout_add(100, pollSerial)

        window = gtk.Window()
        window.connect("destroy", gtk.main_quit)
        window.connect("key-press-event", keyEvent)
        window.fullscreen()
        window.add(image)
        window.show_all()
        gtk.main()

    if __name__ == "__main__":
        main()
```

This example brings together the previous Python examples you've seen in this chapter: keyboard control, reading the serial port, and stretching images to full-screen mode.

This program expects to talk to an Arduino that's running the sketch listed earlier in this chapter (see the section "Determining Direction with the Final Sensor").

This program halts if the serial port is not found. If you want to test the program simply from the keyboard without Arduino connected to the serial port, comment out these lines by inserting # at the beginning of each:

```
        ser = serial.Serial('/dev/ttyUSB1', 9600)
        gobject.timeout_add(100, pollSerial)
```

Be sure to change /dev/ttyUSB1 *to the filename of the serial port your Arduino is connected to.*

Let's have a look at some of the code:

❶ Because the images can be rotated using both the keyboard and messages from the serial port, the go() function handles both.

❷ In the previous example, you might have noticed you get an error if you try to go beyond the last image. This code takes the user back to the beginning if she tries to go past the end, and vice versa. If the user moves backward from the first image (pos<0), the program moves to the last image (pos=last). If the user moves past the last image (last<pos), she returns to the first image. The first index of the array of images is 0. The last cell is the length of the list minus one. For example, in a list with three images, the indexes are 0, 1, and 2.

❸ Calling go() with an argument of 1 will move to the next image; go(-1) moves to the previous image.

❹ Setting timeout to 0 prevents ser.read() from blocking (waiting forever for input from the Arduino). This way, you'll be able to use the keyboard and Arduino for input simultaneously.

Now the program is missing only some eye candy (animations). For now, run it with an Arduino connected (while running the *interactivePaintingSensor* sketch from the section "Determining Direction with the Final Sensor"), and try to choose images by waving your hand in the air.

Animating the Sliding Image

As the icing on the cake, we'll add an animation to the painting. Images will slide to their place. Our program is ready.

All functionality from the previous exercises is included in this version. Images spread to full screen, both the serial port and the keyboard are used for controlling behavior, and images are loaded from a folder. The images can be in many different formats, such as JPG, SVG, GIF, or PNG.

This is also an example of a timed gobject.add_timeout() animation. We ask the timer to call for our function, for example, 10 times a second, and during each call we draw a new animation frame.

```python
#!/usr/bin/env python
# interactivePainting.py - Choose full screen image by waving hand.
# (c) Kimmo Karvinen & Tero Karvinen http://BotBook.com

import gtk, os, serial, gobject

# Global variables

dir="data"
pixbufs=[]
image=None
bg=None
pos=0
ser=None
reel=None
x=0
```

```
w=0
speed=0

# Pixbuf manipulation

def fitRect(thing, box):
    # scale
    scaleY=float(box.height)/thing.height
    scaleX=float(box.width)/thing.width
    scale=min(scaleY, scaleX)
    thing.width=scale*thing.width
    thing.height=scale*thing.height
    # center
    thing.x=box.width/2-thing.width/2
    thing.y=box.height/2-thing.height/2
    return thing

def scaleToBg(pix, bg):
    fit=fitRect(
        gtk.gdk.Rectangle(0,0, pix.get_width(), pix.get_height()),
        gtk.gdk.Rectangle(0,0, bg.get_width(), bg.get_height())
    )
    scaled=pix.scale_simple(fit.width, fit.height, gtk.gdk.INTERP_BILINEAR)
    ret=bg.copy()
    scaled.copy_area(
        src_x=0, src_y=0,
        width=fit.width, height=fit.height,
        dest_pixbuf=ret,
        dest_x=fit.x, dest_y=fit.y
    )
    return ret

def newPix(width, height, color=0x000000ff):
    pix=gtk.gdk.Pixbuf(gtk.gdk.COLORSPACE_RGB, True, 8, width , height)
    pix.fill(color)
    return pix

def catenate(left, right): ❶
    "Return a Pixbuf with 'right' catenated on the right side of 'left'. "
    assert left.get_width()==right.get_width() ❷
    assert left.get_height()==right.get_height()
    reel=newPix(left.get_width()+right.get_width(), left.get_height()) ❸
    left.copy_area( ❹
        src_x=0, src_y=0,
        width=left.get_width(), height=right.get_height(),
        dest_pixbuf=reel,
        dest_x=0, dest_y=0
    )
    right.copy_area(
        src_x=0, src_y=0,
        width=right.get_width(), height=right.get_height(),
        dest_pixbuf=reel,
        dest_x=left.get_width(), dest_y=0
    )
    return reel ❺

def getBox(pix, x, width): ❻
    "Return Pixbuf, a slice of pix, starting at x, given width. "
    buf=newPix(width, pix.get_height())
    pix.copy_area(
```

```
                          src_x=x, src_y=0,
                          width=width, height=pix.get_height(),
                          dest_pixbuf=buf, dest_x=0, dest_y=0)
              return buf

      # File reading

      def loadImages():
          global pixbufs
          for file in os.listdir(dir):
              filePath=os.path.join(dir, file)
              pix=gtk.gdk.pixbuf_new_from_file(filePath)
              pix=scaleToBg(pix, bg)
              pixbufs.append(pix)
              print("Loaded image "+filePath)

      # Controls

      def go(relativePos):
          global pos, reel, x, speed
          last=len(pixbufs)-1
          if pos<0:
              pos=last
          elif pos>last:
              pos=0

          if 0<relativePos:
              print("Next")
              if pos==last:
                  right=0
              else:
                  right=pos+1  ❼
              reel=catenate(pixbufs[pos], pixbufs[right])
              x=0  ❽
              speed=60
          if relativePos<0:
              print("prev")
              if pos==0:
                  left=last
              else:
                  left=pos-1
              reel=catenate(pixbufs[left], pixbufs[pos])
              x=w
              speed=-60
          print("pos == "+str(pos))
          pos+=relativePos  ❾

      def animateSlide():
          global reel, x, speed
          if speed!=0:  ❿
              x+=speed  ⓫
              if x>=w or x<=0:  ⓬
                  speed=0
              print x, reel
              pix=getBox(reel, x, w)  ⓭
              image.set_from_pixbuf(pix)  ⓮
          return True

      def keyEvent(widget, event):
          global pos, image
          key = gtk.gdk.keyval_name(event.keyval)
```

```
        if key=="space" or key=="Page_Down":
            go(1)
        elif key=="b" or key=="Page_Up":
            go(-1)
        elif key=="q" or key=="F5" or key=="ESC":
            gtk.main_quit()
        else:
            print("Key "+key+" was pressed")

    def pollSerial():
        if ser.inWaiting()<=0:
            #print("No data waiting in serial buffer.")
            return True # call again later
        cmd=ser.read(size=1)
        print("Serial port read: \"%s\"" % cmd)
        if cmd=="F":
            go(1)
        elif cmd=="B":
            go(-1)
        return True

    # Main

    def main():
        global bg, image, ser, w

        w=gtk.gdk.screen_width()
        h=gtk.gdk.screen_height()
        window = gtk.Window()
        window.connect("destroy", gtk.main_quit)
        window.connect("key-press-event", keyEvent)
        window.fullscreen()
        bg=newPix(w, h)
        loadImages()
        image=gtk.image_new_from_pixbuf(pixbufs[pos])

        ser = serial.Serial('/dev/ttyUSB0', 9600, timeout=0)
        gobject.timeout_add(100, pollSerial)

        gobject.timeout_add(30, animateSlide) ⑮

        window.add(image)
        window.show_all()
        gtk.main()

    if __name__ == "__main__":
        main()
```

Let's look at the key parts of the code:

❶ Join two images together, side by side, for the animation.

❷ The images to be joined must be exactly the same height and width. If we accidentally try the program with different image sizes, we want to know about this mistake immediately. The assert command interrupts the execution of the program with an error message, if the condition following it is not true. If assert interrupts the execution of the program, it means there is a problem.

❸ Create a new Pixbuf image buffer, onto which both images are copied.

> *Be sure to change /dev/ttyUSB1 to the filename of the serial port your Arduino is connected to.*

This is necessary because with the copy_area() method, we can draw only to the area that already exists within the source image. We use the newly defined newPix() function, and we will store the buffer under the name reel.

❹ Copy the left-side image called left to its place in image reel. We use the copy_area() method from the left-side image. Let's copy starting from the upper-left corner of the image left (0,0) for the whole width and height (get_width()...). Let's set the image "reel" (dest_pixbuf) to the upper-left corner (0, 0). Then copy the right-side image the same way.

❺ Finally, the function returns the image reel, which is a combination of images left and right. This function does not relate to the global function reel of the same name. Here, reel is a local variable, which will become available for the caller as a value returned by the function.

❻ This function takes a slice of the image. It creates a new Pixbuf image buffer with newPix(). Then it copies an area of the requested size. Since the copy_area() method has so many parameters, we have used an option available in Python of marking the name of the parameter in the call. The number of parameters in the copy_area() method is a good reason to separate getBox() into its own function.

❼ This makes a reel for the animation that has the current image on the left and the next picture on the right. The number of the current image is pos, so the current image buffer is pixbufs[pos].

❽ For the animation, the current position is on the left side of the reel, x=0. This sets the update speed as 60 pixels to the right. The animateSlide() function uses these values.

❾ Update the current position. Here, the relativePos parameter is 1, as defined in the go(1) function call.

❿ The image is updated and moved only if speed (speed) has been defined for it.

> animateSlide() *always returns* true, *which allows the timer to call it again no matter what happens within it.*

⓫ Move the left side position forward as defined by speed. When moving the image backward, speed is negative.

⓬ If the program has already moved to another image, this stops the movement. When moving the image backward (to the left), we will face the left side of the image reel, x<=0. When moving the image forward (to the right), we will face the center point of the image reel, x>=w. Then the right side of the image shown will be exactly on the right side of the image reel.

⓭ Pick a screenwide segment from the image reel, beginning at position x. We have previously stored the width of the screen to the variable w. All images have been scaled to be as wide as the screen.

Chapter 5

⑭ Present the Pixbuf named `pix` by copying it to the image widget `image`.

⑮ Set the timer to call `animateSlide()` every 30 milliseconds. This is approximately 33 times a second, which is 33Hz more often than in a TV image:

```
1 s / 30 ms = 1000 ms / 30 ms = 1000/30 = 100/3 = 33.33...
```

GENERATING TEST MESSAGES FROM ARDUINO

If you'd like to mock something up in Arduino without hooking up the sensors, give the following sketch a try. It will send 10 forward (F) messages, 10 backward (B) messages, and then an occasional random movement message:

```
void setup()
{
  Serial.begin(9600);

  // Seed the random number generator
  randomSeed(analogRead(0));

  // Send 10 Fs
  Serial.print("FFFFFFFFFF");

  // Send 10 Bs
  Serial.print("BBBBBBBBBB");
}

void loop()
{
  // Randomly send an F, a B, or nothing.
  if (random(0,10) > 7) {
    Serial.print("F");
  } else {
    if (random(0,10) > 7) {
      Serial.print("B");
    }
  }
  delay(2000);
}
```

Figure 5-24. *The Processing integrated development environment*

Please note the code listed in bold, because it contains important information about configuring Processing to talk to the right serial port.

Connecting Arduino with Processing

Processing (*http://processing.org/*) is an open source language for creating animations, images, and interactive software. You can download the Processing integrated development environment from *http://processing.org/download/*. One of the first things you might notice is that it looks just like Arduino (see Figure 5-24). This is because Arduino and Processing are sister projects that both embrace the goals of simplicity, open source, and accessibility to a wide audience.

Although Processing and Arduino are different programming languages, they share enough similarities that it is easy to switch between them.

Just as with Arduino, Processing programs are known as *sketches*. The following simple sketch reads the serial port and displays the output in the Processing window. To run it, click the leftmost button (Run).

```
import processing.serial.*; ❶

// A serial port that we use to talk to Arduino.
Serial myPort; ❷


// The Processing setup method that's run once
void setup() { ❸

  size(320, 320); // create a window ❹

  // List all the available serial ports: ❺
  println(Serial.list());

  /* On some computers, Arduino will usually be connected to
   the first serial port. If not, change the 0 to the
   correct one (examine the output of the previous line
   of code to figure out which one to use). */
  myPort = new Serial(this, Serial.list()[0], 9600);
}


/* The draw() method is called up to 60 times a second
 unless you change the frame rate of Processing.

 Normally, it is used to update the graphics onscreen,
 but we're just polling the serial port here.
 */
void draw() { ❻

  // Put up a black background.
  background(0);

  // Read the serial port.
  if (myPort.available() > 0) { ❼

    char inByte = myPort.readChar();
    print(inByte); // Displays the character that was read
  }
}
```

Chapter 5

Let's look at the key lines of code in this sketch:

❶ Processing comes with a number of libraries that extend its capabilities. This line loads the serial library, which lets Processing talk to serial ports.

❷ Declare an object named `myPort` of the type `Serial`.

❸ The `setup()` method functions just like the `setup()` method in Arduino: it's called once per sketch.

❹ Create a window of the specified width and height (in pixels).

❺ You'll see something similar to these bold lines in many Processing sketches that talk to the serial port. The first line prints out all the serial ports that Processing can find (see Figure 5-25). On most computers, the Arduino will be connected to the first serial port; this is because most modern computers don't have any built-in serial ports, so the lowest-numbered port is almost always the one Arduino is using. If this is not the case, examine the output in the Processing window and change the 0 on the last bold line to the index of the correct serial port.

Figure 5-25. *Processing listing its serial ports*

❻ The `draw()` function is similar to Arduino's `loop()` in that it's called continuously. Because Processing draws graphics and animations, this function is called a certain number of times per second (60) unless you change the frame rate with the `frameRate()` function.

❼ This block checks to make sure there is some text to read from the serial port. If so, it reads it and prints it out (see Figure 5-26).

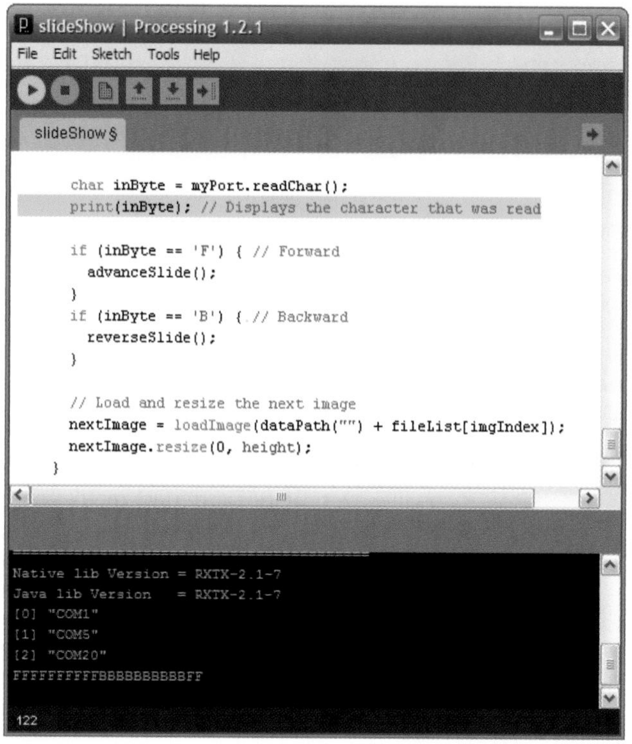

Figure 5-26. *Processing displaying the characters it reads from the Arduino over the serial port*

Processing Code for the Painting

The following Processing program, like our earlier Python program, displays images based on the instructions arriving via the serial port. The images change when a user slides them by waving his hand.

When F is read from the serial port, the images are moved to the right for the width of the painting (screen width). B will likewise move images to the left, for the same number of pixels.

When the user is about to move past the final image, the program moves to the opposite end of the image queue. This will make it appear to the user as if the images are continuing endlessly in both directions.

You must put some images (JPEG, GIF, or PNG) into the sketch's data directory. To find this directory, choose Sketch→Show Sketch Folder. If it does not exist, create a directory, call it data, and put your images in it.

```
// http://BotBook.com
import processing.serial.*;

int slideStep = 75;     // how many pixels to slide in/out ❶

// The current image and the next image to display
PImage currentImage, nextImage; ❷

// The index of the current image.
int imgIndex = 0; ❸

// Keeps track of the horizontal slide position. A negative number
// indicates sliding in from the left.
```

```
int slideOffset; ❹

// All the image files found in this sketch's data/ directory.
String[] fileList; ❺

// A serial port that we use to talk to Arduino.
Serial myPort;

// This class is used to filter the list of files in the data directory
// so that the list includes only images.
class FilterImages implements java.io.FilenameFilter { ❻

  public boolean accept(File dir, String fname) {
    String[] extensions = {".png", ".jpeg", ".gif", ".tga", ".jpg"};

    // Don't accept a file unless it has one of the specified extensions
    for (int i = 0; i < extensions.length; i++) {
      if (fname.toLowerCase().endsWith( extensions[i])) {
        return true;
      }
    }
    return false;
  }
}

// This loads the filenames into the fileList
void loadFileNames() { ❼
  java.io.File dir = new java.io.File(dataPath(""));
  fileList = dir.list(new FilterImages());
}


// The Processing setup method that's run once
void setup() {

  size(screen.width, screen.height); // Go fullscreen

  loadFileNames();    // Load the filenames

  /* This centers images on the screen. To work correctly with
   this mode, we'll be using image coordinates from the center
   of the screen (1/2 of the screen height and width) .
   */
  imageMode(CENTER); ❽

  // Load the current image and resize it.
  currentImage = loadImage(dataPath("") + fileList[0]); ❾
  currentImage.resize(0, height);

  println(Serial.list()); ❿

  myPort = new Serial(this, Serial.list()[0], 9600); ⓫
}

// Go to the next image
void advanceSlide() { ⓬
  imgIndex++; // go to the next image
  if (imgIndex >= fileList.length) { // make sure we're within bounds
    imgIndex = 0;
  }
```

```
    slideOffset = width; // Start sliding in from the right
}

void reverseSlide() {
  imgIndex--; // go to the previous image
  if (imgIndex < 0) { // make sure we're within bounds
    imgIndex = fileList.length - 1;
  }
  slideOffset = width * - 1; // Start sliding in from the left
}

void draw() {

  // Put up a black background and display the current image.
  background(0);
  image(currentImage, width/2, height/2); ⓭

  // Is the image supposed to be sliding?
  if (slideOffset != 0) { ⓮

    // Load the next image at the specified offset.
    image(nextImage, slideOffset + width/2, height/2);
    if (slideOffset > 0) { // Slide from the right (next) ⓯
      slideOffset -= slideStep;
      if (slideOffset < 0) {
        slideOffset = 0;
      }
    }
    if (slideOffset < 0) { // Slide from the left (previous)
      slideOffset += slideStep;
      if (slideOffset > 0) {
        slideOffset = 0;
      }
    }
    if (slideOffset == 0) { ⓰
      currentImage = nextImage;
    }
  }
  else {

    // If we're not sliding, read the serial port.
    if (myPort.available() > 0) {

      char inByte = myPort.readChar();
      print(inByte); // Displays the character that was read

      if (inByte == 'F') { // Forward
        advanceSlide(); ⓱
      }
      if (inByte == 'B') { // Backward
        reverseSlide();
      }

      // Load and resize the next image
      nextImage = loadImage(dataPath("") + fileList[imgIndex]); ⓲
      nextImage.resize(0, height);
    }
  }
}
```

Let's examine the code one piece at a time:

❶ When the user slides to the next or previous image, the program gradually draws the new image over the current one. This variable determines how many pixels at a time the new image will slide in. A value of 1 appears very smooth, but it's slow. You can try changing this value to get something you like.

❷ Processing uses a data type called PImage to represent images. This program uses two of these objects: one to represent the current image, and the other to represent the one that's sliding into view.

❸ This is an index of the list of images, so we know which image we're currently on.

❹ Because the draw() function is called many times a second, we can't waste any time inside it. So we'll be sliding new images into view over a series of calls to draw(). This variable keeps track of where we are in that process. A positive value here indicates images are sliding from the right; a negative value indicates they slide from the left.

❺ This contains the list of all files found in the sketch's *data* directory.

❻ This is a function that's used by the loadFileNames() function to eliminate any files in the *data* directory that aren't images.

❼ Load all the files in the *data* directory into the fileList array.

❽ Tell Processing that it needs to center any images that it draws to the screen. As a result, we won't be using 0,0 as the starting point for drawing images. Instead, we'll use dead center: half of both the height and width.

❾ List all the available serial ports.

❿ On some computers, Arduino will usually be connected to the first serial port. If not, change the 0 to the correct one (examine the output of the previous line of code to figure out which one to use).

⓫ Load the first image, which is at index 0 in the array of files. Because we have only the filenames in this array, we use the dataPath() function to insert the path of the sketch's *data* folder, much as we did with Python and the os.path.join() function.

⓬ Get the sketch ready to move to the next slide by incrementing the image index. Additionally, this section avoids running past the end of the list of images by starting at 0 each time a user tries to go beyond the last image. Finally, it sets the slideOffset to begin sliding in from the rightmost column of pixels onscreen. The next function does the same, but in reverse (decrements the image index, wraps around to the end if the index goes below 0, and prepares slideOffset to slide in from the left).

⓭ Display the current image at the center of the screen.

⓮ If the slideOffset is something other than 0, it means the program is in the middle of sliding the image. The code in this block moves to the next step of the animation.

⑮ This code slides the image in from the right, one step at a time. First it shifts the slideOffset, and then it makes sure it hasn't reached or gone past 0 (if it has gone beyond, this code resets it to 0). The next block does the same for sliding from the left.

⑯ If the slideOffset just reached 0, it's time to swap the images; the current image is replaced by the one we've been sliding in. The sketch will now show the current image until it gets another F or B command from the Arduino.

⑰ If the Processing sketch reads an F from the Arduino, it's time to move forward. Similarly, if it receives a B, it's time to move back.

⑱ Load the next image (the one that's sliding into view) into the nextImage object, so the program can gradually draw it over the current image.

The Finished Painting

Wave your hand in front of the painting, and the frog will change into a landscape. What kind of exhibition could use this type of user interface? In which other Processing programs could you use Arduino's sensors?

You now have an image presentation solution that you can control by waving your hand in front of a computer. If you want to make this into a more permanent device, follow the instructions in the next section.

Creating an Enclosure

Use a metal saw to cut a 25cm piece of PVC tube (Figure 5-27) with a 7.5cm diameter and then smooth the edges with sandpaper (we used P240-grade paper). With a marker, draw a horizontal center line in the pipe; we'll use this as the center point of the ultrasonic sensors. Mark a center point on the tube at 12.5cm. Put the first ultrasonic sensor at that position and trace it on the tube with a marker. Trace the next two sensors the same way, so that their center points are 6.5cm from the center point of the tube (Figure 5-28).

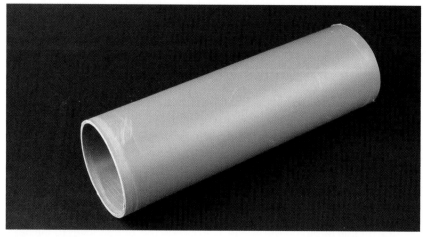

Figure 5-27. A 25cm piece of PVC tube

Chapter 5

Figure 5-28. *The marked ultrasonic sensor positions*

Drill holes with a 16mm bit and smooth their edges with sandpaper (Figure 5-29). Test whether the sensors fit in the holes (Figure 5-30). You might have to trim the holes a bit with the sandpaper, a file, or a mini drill. Clean and dry the surface of the tube. Paint the tube with matte black spray paint (Figure 5-31). (In practice, any regular spray paint works here.) Paint several thin layers to get an even end result.

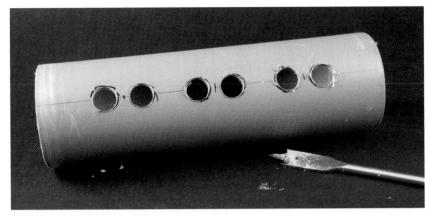

Figure 5-29. *The 16mm holes for the sensors*

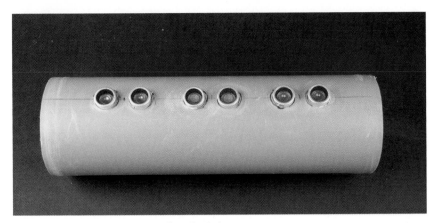

Figure 5-30. *Test that the sensors fit in the holes*

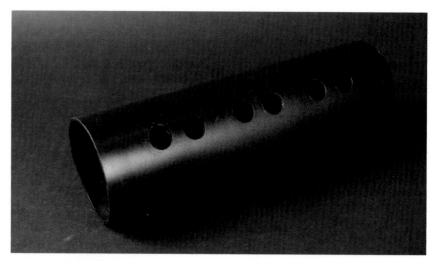

Figure 5-31. *Painted tube*

Attaching the Sensors with Servo Extension Cables

We'll attach the ultrasonic sensors with servo extension cables. Remove the end of the extension cable that fits into the servo connector (Figure 5-32). You can crack and break it with the pliers. This way, you do not have to solder single-strand wires to replace the multistrand ones.

Do not cut the extension cable, because the wires under the plastic have metal ends that fit straight into the prototyping breadboard.

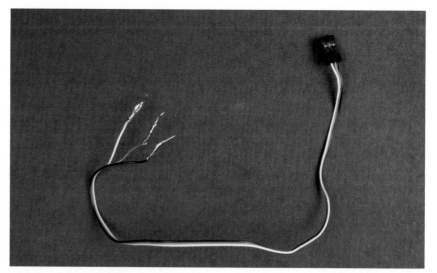

Figure 5-32. *A dismantled extension cable*

Connect a black wire from the Arduino GND pin to the top horizontal line of the prototyping breadboard, and from the +5V pin to the bottom horizontal line. Connect the black and red wires of the extension cable to these rows. Place the wires in different sides of the Arduino to prevent them from accidentally creating a short. Connect the signal wires of the ultrasonic sensors to the same pins used in "Detecting Motion Using Ultrasonic Sensors" (Figures 5-33 and 5-34).

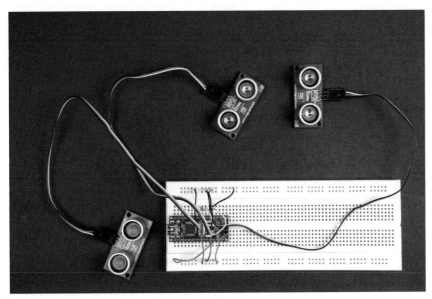

Figure 5-33. *Ultrasonic sensors attached to the prototyping breadboard*

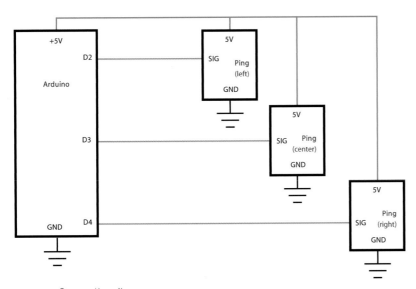

Figure 5-34. *Connection diagram*

Attach the ultrasonic sensors to their holes and push the prototyping breadboard inside the tube.

After the sensors and the electronics are attached, the tube probably doesn't point where you want it to point. You can fix this by gluing something heavy inside the tube to balance it. For example, a strip of bitumen carpet, available from automotive supply stores, can serve this purpose perfectly (Figure 5-35).

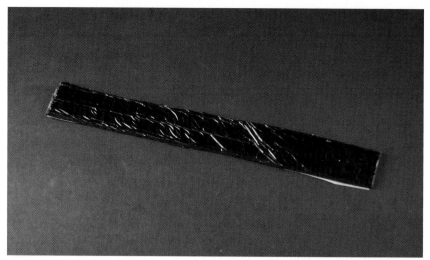

Figure 5-35. *Strip of bitumen carpet for balancing the tube*

Figures 5-36 and 5-37 show the parts placed inside the tube and the completed enclosure, respectively.

Figure 5-36. *Parts in place inside the tube*

Figure 5-37. *Completed enclosure*

Building a Frame

You can use an old, retired laptop computer to build a wall-mounted picture frame. First, turn the laptop around so that the display points in one direction and the keyboard in the opposite direction. In the case of our laptop, this required a small cutting operation for the frame of the computer so that the display cables would reach as far as needed.

Our next challenge was to build a sturdy frame that would handle the weight of the laptop and that you would dare to mount on the wall. You'll find similar projects on the Internet, but many of them have weak supporting frames. Figures 5-38 through 5-41 show a mechanical drawing of the frame and an image sequence of the construction process.

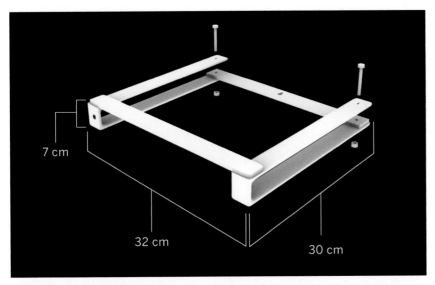

Figure 5-38. *Mechanical design*

Figure 5-39. *Steps 1–4*

1. The old flat iron we chose as the frame material.

2. Sanding down rust and unevenness with an edge sander.

3. The sanded surface.

4. Hammering an arch to the frame with a sledgehammer to fit the portable computer.

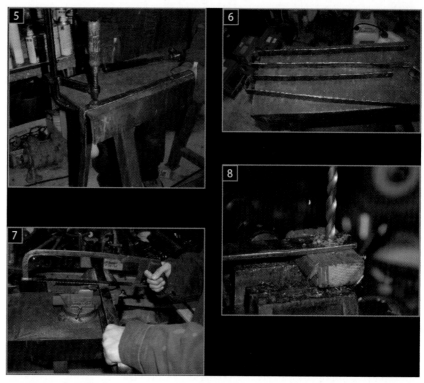

Figure 5-40. *Steps 5–8*

5. Bending the metal in the other direction.

6. The finished bends.

7. Cutting away the excess metal (extra pieces were used for making two metallic beams attached to the frame).

8. Making holes for the upper parts for tightening screws. The laptop will fit tightly in the frame with them. We made a hole in the middle of the upper horizontal beam for wall mounting.

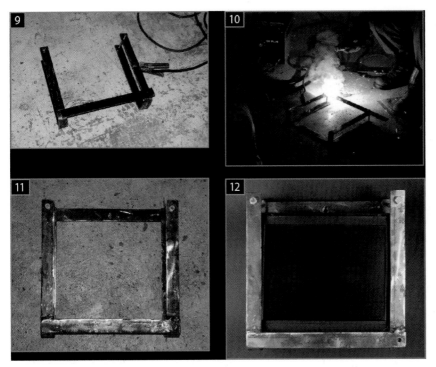

Figure 5-41. *Steps 9–12*

9. The lower cross plate, ready to be welded.

10. Welding the upper cross plate.

11. The finished frame.

12. The laptop inside the frame.

By itself, the frame is quite plain, so as a final step you may want to paint or decorate it to work with your interior design.

Boxing Clock

A round starts. You throw a left jab, followed by a right hook. The crowd is cheering. After a few minutes, the round ends. Saved by the bell! The slowly filling timer shows you how much time you have before the next round. During the break, you remember how you learned to program cell phones while coding the Boxing Clock (Figure 6-10) in this project.

In this project, you will become familiar with cell phone programming, including how to install and run programs on a cell phone. The program you'll create includes the basic components necessary for many other programs. You will learn how to create a graphical user interface, how to draw on the screen, and even how to animate. You will also play MP3 sounds on a cell phone.

We will explain all the code, line by line, but our purpose is not to turn you into a cell phone programming professional. Rather, the goal is to familiarize you with yet another embedded system that can be used in prototyping. As with most programming platforms, the most difficult phase is the installation of the development environment and the first "Hello World" code. After that, it's easy to gather more information and move on to more complex programs.

In a later project, (the Soccer Robot in Chapter 8), you'll use a cell phone to communicate with Arduino. In this chapter, you'll work with a cell phone only to create a round and break timer suitable for boxing matches or even chess boxing (*http://en.wikipedia.org/wiki/Chess_boxing*).

Figure 6-1. *The Boxing Clock on an Android phone*

This chapter covers the Android version of the Boxing Clock. If you'd like the code for Nokia Series 60 and Symbian phones running PyS60 (Python for Series 60/Symbian phones), you can obtain it from http://examples.oreilly.com/0636920010371 *or* http://BotBook.com/. *We're also planning to offer a short ebook for sale that includes a detailed walkthrough of all the Series 60/Symbian examples. For more information on this, check out the book's website or get in touch with us (see "We'd Like to Hear From You" in the Preface).*

What You'll Learn

In this chapter, you'll learn:

- The basics of cell phone programming

- How to install your own programs on a cell phone

- How to install necessary programming tools

Figure 6-2. *Nokia N95 (left) and Google Nexus One (right)*

Tools and Parts

Here are the tools and parts you'll need for this project, shown in Figure 6-2:

- One of the following cell phones:

  — Nokia Series 60–based cell phone, such as the Nokia N95.

  — Android cell phone running Android version 2.2 or newer. We used the Google Nexus One and Sprint EVO 4G, both manufactured by HTC.

- Data cable to connect the cell phone to your computer's USB port.

Android Software Installation

Android is a cell phone operating system developed by Google. It is based on Linux, but most applications are written in Java using the Android SDK provided by Google. Many manufacturers make Android cell phones, including HTC, Samsung, and Motorola.

Programs created with Android are easy to distribute to consumers. For example, you can distribute your own Android application via a web page, or you can put it in the Android Market.

Before learning how to program, you'll need to install several software packages. In addition to Android development tools, you'll need Java and Eclipse, an integrated development environment (IDE). To write applications in Eclipse, you will also need to install an Android extension for it. Eclipse offers a text editor that adds color syntax highlighting to source code; project management capabilities; integrated documentation; and all kinds of other features. You can also extend Eclipse to support many other programming languages in addition to Java.

You'll also need to install Apache Ant, an optional component that will let you compile Java applications, including Android applications, from the command line.

After installation, you'll give the new development environment a try by executing a "Hello World" program, created quickly using an application template supplied with the Android SDK. You'll test the program directly within your computer by using an Android emulator.

When the final Boxing Clock program is ready, you'll transfer the code into a real cell phone.

> *Debugging a program like this can be tricky. See the appendix for information on using Android's logging facility to monitor the program's condition.*

This section provides installation instructions for Ubuntu 10.04, Windows 7, and Mac OS X. If you use another supported operating system, you should be able to adapt the instructions to your situation.

Ubuntu Linux Installation

First, you'll need to install Ant, Java, and Eclipse. Although you can get these files from other sources, it is helpful to use the operating system's built-in package management.

These instructions are specific to Ubuntu Linux, but might work with other Debian-based Linux distributions. You should be able to find these packages for other Linux distributions.

> *If you didn't do so in earlier chapters, switch on the Universe repository, which includes some extra open source programs. Run this command at the Terminal shell prompt:*
>
> ```
> $ sudo software-properties-gtk --enable-component=universe
> ```

Run these commands at the Terminal shell prompt to refresh the available packages and install Ant, Java, and the Eclipse IDE:

```
$ sudo apt-get update
$ sudo apt-get install --yes ant openjdk-6-jdk
$ sudo apt-get install --yes eclipse
```

> *If you have a 64-bit operating system, you will still need the compatibility library, ia32-libs. Trying to install this on 32-bit systems won't harm anything, but the package will install only into a 64-bit operating system:*
>
> ```
> $ sudo apt-get install --yes ia32-libs
> ```

Now it's time to install the Android development tools. Download the installation program from *http://developer.android.com/sdk/index.html* and save the package file in the directory (such as your home directory) where you wish to extract it.

At the Terminal shell prompt, uncompress the package (in geek language, this is known as a *tarball* because it is archived with the *tar* utility):

```
$ tar -xf android-sdk_r09-linux_x86.tgz
```

Replace *09* with the actual version of the SDK (note that the filename may vary by more than the version number in future releases).

This should create a directory called *android-sdk-linux_x86*, although the directory name could change in future releases of the SDK. You can confirm that the package unpacked correctly by running the command `ls android-sdk-linux_x86` and making sure you see several files and directories in there:

```
$ ls -l android-sdk-linux_x86/
total 16
drwxr-xr-x 2 user user 4096 2010-08-30 15:24 add-ons
drwxr-xr-x 2 user user 4096 2010-08-30 15:24 platforms
-rw-r--r-- 1 user user  828 2010-08-30 15:24 SDK Readme.txt
drwxr-xr-x 4 user user 4096 2010-08-28 20:43 tools
```

Windows 7 Installation

You'll need to download and install several packages on Windows.

Java 6 for Windows

Go to *http://www.oracle.com/technetwork/java/javase/downloads/index.html* and download the latest version of the Java Platform, Standard Edition (at the time of this writing, that's Java Platform, Standard Edition, JDK 6 Update 23). If you are given a choice between JDK and JRE, choose JDK. Run the installer.

If you are running 64-bit Windows and choose the 64-bit version of the JDK, be sure to choose the 64-bit version of Eclipse in the next step.

Eclipse

Visit *http://www.eclipse.org/downloads/* and download the latest Eclipse IDE for Java Developers for Windows. Extract the zip file to any location on your hard drive, such as *C:\Eclipse*. You will need to add a shortcut to the Eclipse application manually; you can put it on your desktop, Start menu, or both.

Android SDK

Download the latest Windows version of the Android SDK from *http://developer.android.com/sdk/index.html*. Open the zip file and copy the top-level folder to any location on your hard drive (such as *C:\android-sdk-windows*).

Configure your Path

You should configure your Windows Path environment variable (see Figure 6-3) so that you can run Android tools from the command prompt:

1. Open the Start menu, right-click Computer, and choose Properties.

2. Click the link for Advanced System Settings.

3. Click the Environment Variables button on the Advanced tab.

4. Choose Path from the System Variables list and click Edit.

5. Append the following to the end of the current value (do not erase any of the existing values): `;C:\android-sdk-windows\`. (If you put the SDK somewhere else, use that location here instead.)

6. Close any open command prompt windows and reopen them.

> *Be sure you are logged in as a user with administrative privileges. If you're not, you'll need to have the username and password of an admin user handy, because the installers might prompt you for both when you try to install the software. Once you are finished with the administrative tasks, log in again as a normal, nonadministrative user.*

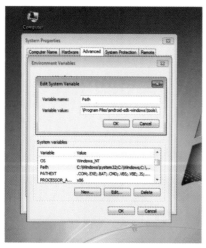

Figure 6-3. *Adding the folder into the Windows Path variable*

Mac OS X Installation

You'll need to download and install two packages on Mac OS X.

Eclipse

Visit *http://www.eclipse.org/downloads/* and download the latest Eclipse IDE for Java Developers for Mac OS X. Extract the *tar.gz* file by double-clicking it, and copy the Eclipse folder to any location on your hard drive, such as */Applications/Eclipse*.

Android SDK

Download the latest Mac OS X version of the Android SDK from *http://developer.android.com/sdk/index.html*. Extract the zip file by double-clicking it, and copy the *android-sdk-mac_** directory to any location on your hard drive (such as */Users/yourname/android-sdk-mac_x86*).

Configuring the Android SDK

Run the Android SDK and AVD manager. On Linux and Mac OS X, you can run it by typing the path to the file. If you've installed the Android SDK in your home directory, type this command at a Terminal shell prompt (the $ is the shell prompt itself; type everything after it):

```
$ ~/android-sdk-*/tools/android
```

On Windows, type **android** at a command prompt or from the Run dialog (Windows-R). Press Enter to run it.

> *If you are using a non-US keyboard, the ~ key may be difficult to access. You can also type* **$HOME/android-sdk-*/tools/android**.

Next, the Android SDK and AVD Manager appears. You should install a version of the Android SDK that targets the broadest number of phones. Android maintains a list of current distributions at *http://developer.android.com/resources/dashboard/platform-versions.html*, which you can use to guide your decision. At the time of this writing, 86% of the devices were running Android 2.1 or later, so we use Android 2.1 in this section.

Let's install Android SDK 2.1 API level 7 and its instructions and example programs. Click the tab labeled "Available packages," and check the box next to Android Repository. After a moment, a list of available tools will open, as shown in Figure 6-4. Select the Android SDK Platform tools as well as the SDK for the version you want to use, such as SDK Platform Android 2.1, API 7. You might also want to select the samples for that version of the SDK.

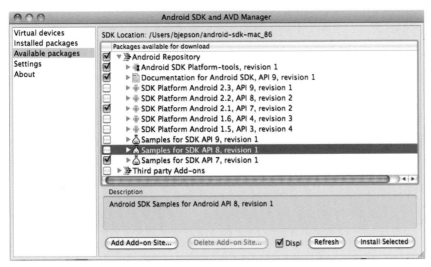

Figure 6-4. *Downloading SDK components*

If you are running Windows, you must also select the USB Driver Package (it is not needed on Linux or Mac OS X). This driver provides support for several phones:

- *T-Mobile G1 (also released as the Google ADP1)*
- *T-Mobile myTouch 3G (also released as the Google Ion)*
- *Verizon Droid*
- *Nexus One*

If you have a similar device on another carrier, the driver might work with them. If not, you will need to obtain a driver from the cell phone carrier or manufacturer. See http://developer.android.com/sdk/win-usb.html *for more information.*

Click Install Selected, and follow the instructions on the screen that appears. You will need to accept all of the license agreements before you can install the packages. When you're done with the installation, you can quit the Android SDK and AVD Manager.

Installing the Android Plug-in for Eclipse

Launch Eclipse. The first time you run it, Eclipse will ask you where to store all your projects. Choose a folder you prefer. If you don't want Eclipse to ask you every time you run it, select "Use this as the default and do not ask again."

The Eclipse welcome screen appears first (Figure 6-5). You can dismiss it and reveal the rest of the Eclipse IDE by clicking the close (X) button on the right of the welcome screen's tab at the top left.

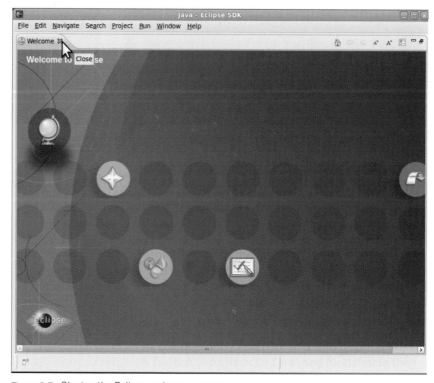

Figure 6-5. *Closing the Eclipse welcome screen*

Chapter 6

Now you can see the Eclipse main screen (Figure 6-6). This is the IDE you will use to develop Android applications.

Now you need to install the Android plug-in:

1. Choose Help→Install New Software, as shown in Figure 6-6.

2. Type **https://dl-ssl.google.com/android/eclipse/** in the "Work with" field and press Enter.

3. You will see a Developer Tools option in the list. Check the box next to it and click the triangle next to the checkbox to expand the view. Confirm that Android DDMS, Android Developer Tools, and Android Hierarchy Viewer are selected, as shown in Figure 6-7, and press Next.

Figure 6-6. *Installing new software via the Help menu*

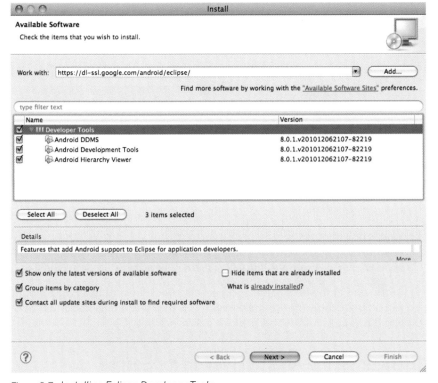

Figure 6-7. *Installing Eclipse Developer Tools*

4. When the install wizard shows a list of the components to be installed, click Next.

5. You must accept the terms of the licenses to proceed, as shown in Figure 6-8. Although they look long and tiresome, these are open source licenses that are actually pretty exciting as far as software licenses go, in that they permit the software to be modified and redistributed under a given set of conditions. For example, as of this writing, the two Android packages used a combination of the Apache 2.0, BSD, and Eclipse plug-in licenses.

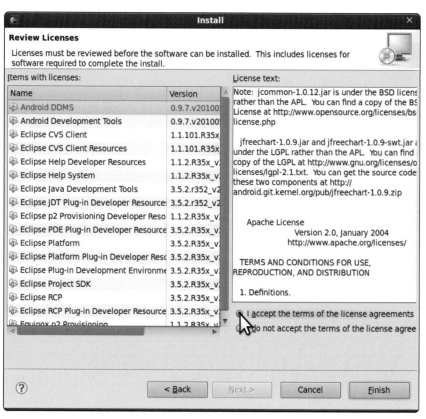

Figure 6-8. *Long list of licenses*

Figure 6-9. *A warning related to unsigned packages*

6. Wait a few minutes for the packages to download. Partway through the download, the program will ask for permission to install unsigned packages, as shown in Figure 6-9. You will need to permit this (click OK) to proceed.

7. At the end of the installation, Eclipse will offer to restart itself. You should restart your computer to make sure all of the new components are activated properly.

Configuring the Android Plug-in for Eclipse

After Eclipse restarts, you need to configure the Android plug-in:

1. On the Mac, click Eclipse→Preferences. On Windows or Linux, choose Window→Preferences.

2. Click Android from the list on the left. The first time you do this, you'll see a dialog box with a welcome message that gives you the option to send usage statistics to Google. Make your selection and click Proceed.

3. Click the Browse button to the right of the SDK Location field (see Figure 6-10) and navigate to the directory where you installed the Android SDK earlier. Select that directory and click OK, as shown in Figure 6-11.

4. Click OK to close the Preferences dialog.

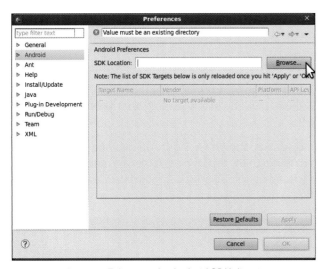

Figure 6-10. *Pointing Eclipse to the Android SDK directory*

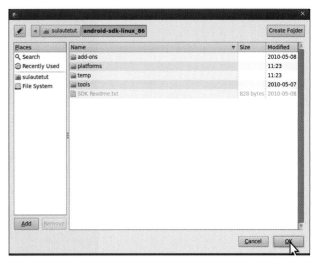

Figure 6-11. *Locating the Android SDK in the home directory where you downloaded it*

Creating a Boxing Clock in Android

Before we start writing the Boxing Clock in Android, let's make sure that the most bare-bones "Hello World" project compiles and runs.

Beginning with "Hello World"

Select File→New→Project, and open the Android section from the dialog that appears (Figure 6-12). Choose Android Project and click Next.

The New Android Project appears. Fill in the information as follows, shown in Figure 6-13.

1. Enter ChessBoxing in the "Project name" field. This will also determine the directory name that's created under your Eclipse workspace folder.

2. Under Build Target, choose the version of Android, such as 2.1 (API level 7), that you want to set as the minimum Android version this app will run under.

3. Under Properties, enter Chess Boxing in the "Application name" field. This is the name that's displayed when the program is running.

4. Next, specify the package name in the traditional Java style by writing your own domain name backward, with the name of the program in the end. For example, our domain is `sulautetut.fi`, so we will start with `fi.sulautetut`. We used `android.chessboxing` for the program name, so the package name we used was `fi.sulautetut.android.chess boxing`. We use this name throughout the chapter, so it will be easier if you use our name. If you replace this with your own name, be sure to use that name wherever you see `fi.sulautetut.android.chessboxing`.

Figure 6-12. *Starting a new Android project with a wizard*

Figure 6-13. *New project info*

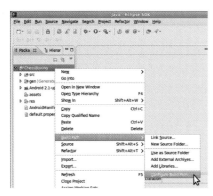

Figure 6-14. *Configuring the Build Path for the Android version you're using to compile the project*

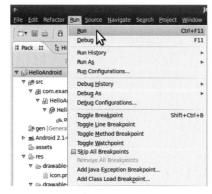

Figure 6-16. *Running the project*

5. Check the Create Activity box to define a class to describe the first (and so far the only) mode of the program. Name this activity ChessBoxing. Since activities are classes, we follow Java's CamelCase naming convention by using an initial capital for each word in the name (no spaces or underscores).

6. Under Min SDK Version, pick the same level (7) that you picked for the build target.

7. Right-click the project (as shown in Figure 6-14), and choose Build →Configure Build Path. Go to the tab labeled "Order and Export" and select the Android library (for example, "Android 2.1-update1," as shown in Figure 6-15). Click OK.

Figure 6-15. *Choosing the Android version you prefer*

8. Run the program by clicking the green Play button in the toolbar or by choosing Run→Run (Ctrl-F11), as shown in Figure 6-16.

9. Select Android Application from the dialog that appears and click OK.

If you've never defined an emulator image to run your programs, and if you don't have a phone connected in debugging mode, Eclipse will warn you that it couldn't find a device or emulator to run under and ask, "Do you wish to add new Android Virtual Device?" Click Yes to begin creating a new emulator image.

The Android SDK and AVD Manager appears (this is the same program we used earlier for loading the SDK):

1. Click New.

2. Select your installed Android version as the target (e.g., 2.1, API 7). Give it a descriptive name, such as "seven."

3. Leave everything else at its default and click Create AVD, as shown in Figure 6-17.

4. Close the Android SDK and AVD Manager.

Now Eclipse will ask you which virtual device you would prefer to use. Click the "Launch a new Android Virtual Device" option, click Refresh (as shown in Figure 6-18), and choose the virtual device "seven" you just created.

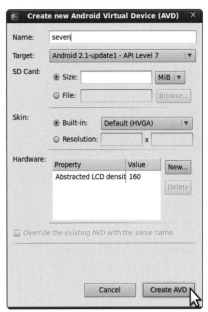

Figure 6-17. *Creating a new emulator*

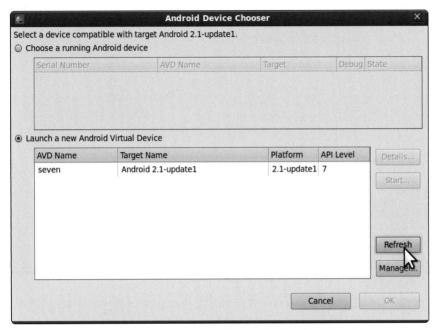

Figure 6-18. *Choosing the emulator to be used (remember to refresh the view)*

The Android emulator will take some time to start, but while you are waiting you can admire the Android logo (Figure 6-19). In our system, starting the emulator took several minutes, so it might be best not to close it until you're done working on the program. Once the emulator is up and running, you can deploy new versions of your program into it by clicking Run.

Finally, the "Hello World" program starts. The title bar shows the name of the program (Chess Boxing), and the screen displays the text "Hello World, Chess-Boxing!", as shown in Figure 6-20.

That one simple Hello World program took a lot of tweaking. Using this program as a starting point, we'll modify it for the next couple of examples in this chapter so you won't have to go through all the steps until later in the chapter (and even then, you won't have to go through *all* of them).

Creating a User Interface

First, we'll use the ChessBoxing project we just created as the basis for a simple user interface that displays text to the user.

Figure 6-19. *The emulator starting*

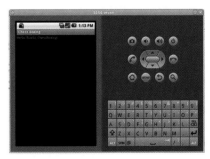

Figure 6-20. *"Hello World" in the emulator, indicating a successful installation*

Within Android, you can't print straight to the console with commands familiar to Java programmers such as `System.out.println`. Instead, we'll create a text box directly in the graphical user interface and print to it:

```
// bca01helloGui - Create a simple graphical user interface
// (c) Kimmo Karvinen & Tero Karvinen http://BotBook.com

package fi.sulautetut.android.chessboxing; ❶

import android.app.Activity; ❷
import android.os.Bundle;
import android.widget.TextView;

public class ChessBoxing extends Activity { ❸
    @Override
    public void onCreate(Bundle savedInstanceState) { ❹
        super.onCreate(savedInstanceState);
        TextView tv = new TextView(this); ❺
        tv.append("Welcome to BotBook.com Chess Boxing!"); ❻
        setContentView(tv); ❼
    }
}
```

Before we run this code, let's go over it:

❶ Declare which *package* the program belongs to. A package will allow you to group together different Java files if needed. If you entered a different reverse domain name and app name when you created the project in the "Beginning with Hello World" section, replace `fi.sulautetut.android` `.chessboxing` with the name you used.

❷ Import the classes needed to support the Android APIs you need. It can be hard to remember which packages a certain API comes from, so you can just write your code and press Ctrl-Shift-O on Linux or Windows (or Command-Shift-O on the Mac) to update the list of imports.

❸ Define a class. Use the same class name you chose when you set up the project, because this project will look for an Activity named ChessBoxing. Just as with the code that was generated when you created a new project, your `ChessBoxing` class extends the class `Activity`.

This class is your custom version that controls what happens when a new ChessBoxing activity is created.

❹ Override the `onCreate()` method that is defined in Android's `Activity` class. If you examine the code that was generated at the creation of the project, you'll see the `@override` *decorator*; leave that in here.

At the start of the method, call the parent class's `onCreate()` method with `super.onCreate()`. ChessBoxing is the main activity of our program. It will start automatically when the program starts.

Main activity methods (functions) resemble the Arduino functions `setup()` and `loop()`, even though Android and Arduino are otherwise quite different systems. `ChessBoxing.onCreate()` will run once at the beginning of the program, just like Arduino's `setup()`. Later, you'll create a timer to run your own function evenly five times per second, just like the Arduino's `loop()`.

❺ Make a user interface within the onCreate() method. This differs from the program generated earlier, in which the user interface was defined using an XML file named *main.xml*. That file will continue to exist in the project, but you will no longer be using it.

Make a new text view (TextView) and assign it the name tv.

❻ We will add text to the TextView using its append() method.

❼ Finally, use the setContentView() method to display the view on the screen. A *view* in Android is a rectangular area of the cell phone screen. In our case, TextView tv can manage the entire screen of our program. A view is an object you can control using different methods.

Now you've seen how to create a user interface and display text straight from program code. You also know something about the structure of an Android program. Let's run the program.

The ChessBoxing app has just one Java file, *ChessBoxing.java*. Find this file in Eclipse by going to the list on the left side of the screen and opening the src group under the ChessBoxing project. Next, open the package name (such as fi.sulautetut.android.chessboxing) and double-click *ChessBoxing.java* to open the file in the editor (see Figure 6-21). Replace all the code in *ChessBoxing .java* with the code in the previous listing. Click Run or press F11 to run ChessBoxing.

> You might be wondering how to quit the program. There is no way to do this within Android. If the user presses the home button and runs a different program, this program will remain in the background. However, if there is not enough memory available to run multiple programs, the operating system might terminate those in the background.

Figure 6-21. *Cutting and pasting the program code over the Java file*

Figure 6-22. *The custom graphical user interface within the emulator*

When ChessBoxing.onCreate() runs, a text box displays "Welcome to BotBook .com Chess Boxing!" on the screen, as shown in Figure 6-22.

The program does not override any other Activity class methods, which means that it does not respond to any events, such as user input.

Using a Timer for the Main Loop

Repeat this, repeat this, repeat this. Animations, games, and many other programs are all based around a main loop. In fact, you've been programming Arduino this way: most of your execution time has been spent within the `loop()` function.

Computers and cell phones usually execute many programs at the same time, so we can't hog all the processing time as we do on Arduino; we have to share resources with other programs. Therefore, we will execute our main loop by triggering it periodically—for example, five times a second. That way, when the operating system needs to let another task do something, our program will know how to behave.

The following program creates a timer that prints "Bling!" on the screen every 10 seconds. In a later example, we will use the same basic approach, but it will update an animation instead of printing text.

```
// bca02chessBoxing - Timed events
// (c) Kimmo Karvinen & Tero Karvinen http://BotBook.com
package fi.sulautetut.android.chessboxing;

import android.app.Activity;
import android.os.Bundle;
import android.os.Handler;
import android.widget.TextView;

public class ChessBoxing extends Activity {
    private Handler handler; ❶
    private TextView tv; ❷

    @Override
    public void onCreate(Bundle savedInstanceState) { ❸
        super.onCreate(savedInstanceState);
        tv = new TextView(this);
        tv.append("Welcome to BotBook.com Chess Boxing! ");
        setContentView(tv);

        handler = new Handler(); ❹
        handler.removeCallbacks(update); ❺
        handler.postDelayed(update, 50); ❻
    }

    private Runnable update = new Runnable() { ❼
            public void run() { ❽
                tv.append("Bling! ");
                handler.removeCallbacks(update);
                handler.postDelayed(this, 10*1000); ❾
            }
    };
}
```

Copy the contents of this program into *ChessBoxing.java* and run it. Figure 6-23 shows the output.

Figure 6-23. *Chess Boxing program printing "Bling!" in 10-second intervals*

Chapter 6

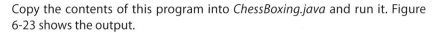

This program has many similarities with the previous example. The only new feature in the code is the timer. You will define a new class variable of the type `Handler` for the entire `ChessBoxing` object. In the beginning of the program, within the `onCreate()` method, you will store a new `Handler` object and ask the timer to call the `update()` function, which is essentially your main loop.

❶ Anything that appears in this part of the class declaration will be visible to all methods within the same class. For example, here the `Handler` variable is usable within the `onCreate()` method as well as in the `update()` method.

   We've specified the visibility of the variables as `private`. Within Java, any objects, methods, and variables marked as `public` are meant for use by other classes. If you are not building an API for other classes, you might as well set the visibility to `private`.

   The `Handler` class will define the timers, and we will create a `Handler` type of a variable called `handler`.

❷ The previous program wrote to a text view, `tv`, from only one method, `onCreate()`. Now, the `update()` method must also access the text view, so the program must declare `tv` in the beginning of the class.

❸ In the beginning of the program, call the default activity's `onCreate()` method.

❹ Create a new `Handler` object as the variable `handler`. A `Handler` class constructor does not take any parameters.

❺ When you're setting up a new timer, it is useful to remove old timers to ensure that they do not accrue on top of each other.

   We have to tell the `removeCallBacks()` method which timers are removed. Since this program only times update objects, the program gives `update` as its parameter.

❻ Set the first timer, asking the timer handler to call our function after 50 milliseconds. Time is defined in milliseconds (ms), one thousandth of a second.

   Android's `Handler` class is designed in such a manner that the function it invokes is called `update.run()`. We will define the `update` variable itself as a `Runnable` class object, which means we'll have to define the `run()` method.

   Since a millisecond is one thousandth of a second, 50ms equals 0.050s, or one-twentieth of a second (50ms / 1,000ms = 5/100 = 1/20). A human being can't really perceive a time period this short.

❼ Create an `update` object in a special way that allows us to specify the class definition and instantiate it all at once. (Later, you will see a more traditional example of defining classes.) Here, we define a new object of type `Runnable` that's visible to the rest of the class.

❽ The content of our own `run()` method runs the task that the timer invokes. Here, the program just prints "Bling!" to the text view, but in a game or animation, all the other main loop code would go here.

Finally, call the timer again. Just as we did when creating the first timer, remove all existing timers to prevent them from accumulating.

⑨ Finally, request to call the update.run() again after 10 seconds. Time is given in milliseconds. To make it easier to read, it is helpful to specify the time as a multiplication (10 seconds * 1,000 milliseconds).

Now the program will continue forever—or at least, until the phone is restarted (or unless the program runs unused in the background for a time and the OS decides to terminate it).

Adding Sound to the Boxing Clock

When you're boxing, it can be hard to check your cell phone to see if the round has ended. It's time to add sounds to the Boxing Clock to make notification easier.

The first version of the program will simply play back an MP3. Create a new project just as you did earlier in the chapter (close any open files from other projects first). From within Eclipse, choose File→New: Project→Android project. Set the properties shown in Table 6-1 and click Finish.

Table 6-1. Project settings for the MP3 playback example

Setting	Value
Project name	HelloMp3
Build Target	2.1
Application name	Hello mp3!
Package name	fi.sulautetut.android.hellomp3
Create Activity	HelloMp3
Min SDK Version	7

Only lowercase letters, numbers, and underscores can be used in the filenames (if you're familiar with regular expressions, this class is represented as [a-z0-9_]).

Run the program to make sure it works. When prompted to specify how to run it, choose Android Application. Now you can customize it to play an MP3.

First, add the *fightsound.mp3* MP3 file (download it from *http://examples.oreilly .com/0636920010371*) into the project resources. Find the *res* folder in the left side of the Eclipse window, create a new folder (right-click on *res*, choose New→Folder), and name the new folder *raw*. Drag *fightsound.mp3* into this folder. When prompted to specify how to copy the file, choose Copy Files.

This MP3 file will not appear on the cell phone's filesystem. Instead, the Eclipse project will automatically create and import a class named R that exposes resources to your code.

Open up the main activity *HelloMp3.java* (it's under *src/fi.sulautetut.android .hellomp3*). Add the sound handling in the end of its onCreate() method like so:

```
// bca03helloMp3 - Play a sound
// (c) Kimmo Karvinen & Tero Karvinen http://BotBook.com

package fi.sulautetut.android.hellomp3;
```

Chapter 6

```
import android.app.Activity;
import android.media.MediaPlayer; ❶
import android.os.Bundle;

public class HelloMp3 extends Activity {
    @Override
    public void onCreate(Bundle savedInstanceState) {
        super.onCreate(savedInstanceState);
        MediaPlayer fightSound = MediaPlayer.create(this, R.raw.fightsound); ❷
        fightSound.start(); ❸
        setContentView(R.layout.main); ❹
    }
}
```

Let's break down this code:

❶ If you want to try out an interesting feature of the Eclipse IDE, remove this line from the code. You can add it back in automatically in Step 2.

❷ Load the sound from the R object and play it. First, create a new variable of the MediaPlayer type and choose fightSound as its name. Instead of a regular constructor, call the MediaPlayer.create() *factory method*. The first parameter is its context (this, referring to the current class) and the second is the fully qualified name of the resource (fightsound).

If you'd like to see the power of Eclipse, try deleting R.raw.fightsound from the code. Then start typing R., and Eclipse will give you a list of choices (if it does not, press Ctrl-Space). Double-click raw, type a period (.), and choose fightsound from the next list of choices (note that there is no filename extension, such as *.mp3*).

If you deleted the import line back in Step 1, press Ctrl-Shift-O (Linux and Windows) or Command-Shift-O (Mac) and watch it reappear.

❸ The Sound object's method start() plays the sound.

❹ This line establishes a full-screen view (see Figure 6-24) that is defined in an XML file, *res/layout/main.xml* (this file was generated when you created the project). You won't use this in later projects (you'll just create user interfaces with code).

Now you can play sounds.

Figure 6-24. *HelloMp3 playing the MP3 file*

This program plays the sound only once. In fact, if you try to run the program again, Eclipse might inform you that it's already running ("Activity not started, its current task has been brought to the front"). If you'd like to make sure it restarts each time, choose Run→Debug instead of Run→Run.

Creating a Ringing Boxing Clock

Now let's make the Boxing Clock play a sound every 10 seconds. (If you are eager to get to actual practice, you can change this to a full minute.) A ringing Boxing Clock can easily be used in sweatier games than chess. This project combines the timer and MP3 playback features you have already learned.

From within Eclipse, choose File→New: Project→Android project (close any other source code files first). Set the properties shown in Table 6-2 and click Finish.

Table 6-2. Ringing clock project settings

Setting	Value
Project name	LoudBoxing
Build Target	2.1
Application name	Boxing Clock
Package name	fi.sulautetut.android.loudboxing
Create Activity	LoudBoxing
Min SDK Version	7

As in the previous example, create a new folder called *raw* under *res*, and drag *fightsound.mp3* into it.

```java
// bca04loudBoxing - Signal rounds with a gong sound.
// (c) Kimmo Karvinen & Tero Karvinen http://BotBook.com

package fi.sulautetut.android.loudboxing;

import android.app.Activity;
import android.media.MediaPlayer;
import android.os.Bundle;
import android.os.Handler;
import android.widget.TextView;

public class LoudBoxing extends Activity {
        private Handler handler;
        private TextView tv;
        MediaPlayer fightSound;

    @Override
    public void onCreate(Bundle savedInstanceState) {
        super.onCreate(savedInstanceState);
        tv = new TextView(this);
        tv.append("Welcome to BotBook.com Chess Boxing! ");
        setContentView(tv);

        fightSound= MediaPlayer.create(this, R.raw.fightsound ); ❶

        handler = new Handler();
        handler.removeCallbacks(update);
        handler.postDelayed(update, 50);
    }

        private Runnable update = new Runnable() {
                public void run() {
                        fightSound.start(); ❷
                        tv.append("Bling! ");
                        handler.removeCallbacks(update);
                        handler.postDelayed(this, 10*1000); ❸
                }
        };
    }
```

Here's a closer look at the code:

❶   As in the previous example, create a `MediaPlayer` object for the fight sound.

❷ Play the sound from within the main loop.

❸ Repeat the action, using the same technique used in "Using a Timer for the Main Loop."

Now your clock rings once every 10 seconds (Figure 6-25). Go jump rope or play a few rounds of chess; you deserve a little break.

Figure 6-25. *LoudBoxing dinging loudly every 10 seconds*

> Even if you tap the home button, you'll still hear the sound in the background. If you want to stop the program, go to the home screen, tap the Menu button, choose Applications→Manage Applications, and select Boxing Clock. Scroll to the bottom of the screen that appears and tap Force Stop.

Setting Separate Rounds and Breaks

Now we'll set round and break times to be different lengths for cases when you might want to practice for two or three minutes and then have a one-minute break.

You can either create a new project or modify the previous one. This program code uses different sounds for a round and a break, so in addition to *fightsound .mp3*, you'll need to copy *breaksound.mp3* into *res/raw*.

```
// bca05roundAndBreak - Round and break with different length and sound.
// (c) Kimmo Karvinen & Tero Karvinen http://BotBook.com

package fi.sulautetut.android.loudboxing;

import android.app.Activity;
import android.media.MediaPlayer;
import android.os.Bundle;
import android.os.Handler;
import android.os.SystemClock;
import android.widget.TextView;

public class LoudBoxing extends Activity {
        private Handler handler;
        private TextView tv;
        private MediaPlayer fightSound;
        private MediaPlayer breakSound; ❶
        private boolean fight=false; ❷
        private long pieStarted; // round or break, ms ❸
        private long pieEnds;
        private long fightLen=10*1000; // ms ❹
        private long breakLen=5*1000;

        @Override
        public void onCreate(Bundle savedInstanceState) {
                super.onCreate(savedInstanceState);
                tv = new TextView(this);
                tv.append("Welcome to BotBook.com Chess Boxing! ");
                setContentView(tv);

                fightSound= MediaPlayer.create(this, R.raw.fightsound );
                breakSound= MediaPlayer.create(this, R.raw.breaksound );
```

```
                                       handler = new Handler();
                                       handler.removeCallbacks(update);
                                       handler.postDelayed(update, 50);
                              }

                     private Runnable update = new Runnable() {
                              public void run() { ❺
                                       if (pieEnds<SystemClock.uptimeMillis()) { ❻
                                                fight=!fight; ❼
                                                pieStarted=SystemClock.uptimeMillis(); ❽
                                                if (fight) { // round starts ❾
                                                         pieEnds=pieStarted+fightLen; ❿
                                                         fightSound.start(); ⓫
                                                         tv.setText("Fight! ");
                                                } else { ⓬
                                                         pieEnds=pieStarted+breakLen;
                                                         breakSound.start();
                                                         tv.setText("Break. ");
                                                }
                                       }
                                       handler.removeCallbacks(update);
                                       handler.postDelayed(this, 1000);
                              }
                     };
            }
```

Let's examine this code:

❶ Many new attributes have been added to the LoudBoxing activity. This program defines them for the whole class so that it can use them in both onCreate() and update.run().

Create MediaPlayer objects for fightSound and breakSound. (We can't name the break sound break, because it is a reserved word used in Java.)

❷ Define fight with an initial value of false, indicating we have not yet started the fight.

❸ Create variables for the beginning and end times of the timer, which the program will later draw as a slowly filling pie diagram.

❹ If the system is running many other processes, the timer's times may not be very accurate, so it's better to use a real clock. SystemClock.uptimeMillis() tells how much time has passed since the cell phone was powered on. Because the time is presented in milliseconds (1ms = 1/1,000s), the number could be quite large. The type of number is a long integer (long), not just a normal (short) integer.

❺ Invoke the main loop once per second. When we add animation in an upcoming example, we can change update.run() to run many times per second.

❻ If this moment (SystemClock.uptimeMillis()) is longer than the end of the previous round or break (pieEnds), start a new pie.

❼ After a round there will be a break, and after a break there will be a round. So, toggle the fight from true to false (or vice versa if it is already false).

❽   Store the pie's start time.

❾   If the `fight` variable is `true`, the round is just starting.

❿   Set the present moment (the start of the round) plus the length of a round as the new end time. We could have written the length of the round here also, but instead of magical numbers, it's best to use named variables. (If you want three-minute rounds, you can change that in the beginning of the code rather than down here in the guts.)

⓫   Play the start sound and display the text "Fight!" on the screen.

⓬   In the `else` loop (`fight==false`), the break is just starting. Then the break sound (ding ding) plays and the screen displays "Break." The next trip around the pie is set to take place within the length of the break.

Now you have carefully measured rounds and breaks with different lengths (Figure 6-26). Before you go to the mat to test your skills against a friend, you might want to set the length of the fight to a minute, and the length of the break to two minutes. Good luck with the challenge!

Figure 6-26. *Separate rounds and breaks*

Drawing Graphics with Custom Views

Graphics are created using custom views. As discussed in "Creating a User Interface," a *view* is a rectangular area on a cell phone display. This example expands the built-in `View` class and overrides its `onDraw()` method to draw custom graphics.

To begin, close any open projects or source code files, and create a new project with the settings shown in Table 6-3.

Table 6-3. *Custom view project settings*

Setting	Value
Project name	GreenColor
Build Target	2.1
Application name	Green Color
Package name	fi.sulautetut.android.greencolor
Create Activity	GreenColor
Min SDK Version	7

Replace the contents of the *GreenColor.java* file with the following code and run it. You will see a green screen.

```
// bca06customView - Build custom view with green background
// (c) Kimmo Karvinen & Tero Karvinen http://BotBook.com

package fi.sulautetut.android.greencolor;

import android.app.Activity;
import android.content.Context;
import android.graphics.Canvas;
import android.graphics.Color;
import android.os.Bundle;
import android.view.View;
```

```
public class GreenColor extends Activity {
    @Override
    public void onCreate(Bundle savedInstanceState) { ❶
        super.onCreate(savedInstanceState);
        PieView tPie = new PieView(this); ❷
        setContentView(tPie); ❸
    }

    public class PieView extends View { ❹

        public PieView(Context context) { ❺
                super(context);
        }

        public void onDraw(Canvas canvas) {
                canvas.drawColor(Color.GREEN); ❻
        }
    }
}
```

Let's take a look at the code:

❶ As in other examples, the onCreate() method overrides the parent class's (Activity) onCreate() method and immediately calls the parent class implementation through the user of the super object.

❷ Create a new object, tPie, which belongs to the PieView class defined in Step 4.

❸ The setContentView() method uses the custom view to set the contents of the screen.

❹ In short programs, it is helpful to write classes within each other so you don't have to create new files for them. Here, the PieView class is written within the GreenColor class.

❺ This custom view extends the View class, overriding the constructor and onDraw. This constructor does the bare minimum and chains to the constructor of the parent class.

❻ The meatiest part of the view is the onDraw() method, which runs whenever the view must be drawn. You can request a refresh by marking the view as invalid using the invalidate() method. In onDraw(), the program just fills the whole view with a green color.

When you run this program, Android creates its initial activity by invoking GreenColor.onCreate(). A new PieView object is created in it, because the constructor PieView.PieView() runs from within the activity's constructor. When the new PieView object is made visible using the Activity.set ContentView() method, the custom view's PieView.onDraw() runs and a green color fills the view, as shown in Figure 6-27.

Figure 6-27. A custom view, the first step toward drawing an animation

Animating the Pie

Next, the Boxing Clock will show, via the pie diagram, how much time is left. Then you can shout on the side of the field like a real coach, "Now put everything in the game; you're down to the last seconds!"

In this project, you'll learn how to make simple animations. The easiest way to create animations is to combine a custom view and main loop created with a handler, both of which you've seen in earlier projects in this chapter.

Create a new project with the settings shown in Table 6-4.

Table 6-4. *Animated boxing project settings*

Setting	Value
Project name	AnimatedBoxing
Build Target	2.1
Application name	Animated Boxing
Package name	fi.sulautetut.android.animatedboxing
Create Activity	AnimatedBoxing
Min SDK Version	7

Replace *AnimatedBoxing.java* with the following code and run it; you'll see a white pie graphic filling the screen:

```
// bca07animatedBoxing - Animated pie shows time left.
// (c) Kimmo Karvinen & Tero Karvinen http://BotBook.com

package fi.sulautetut.android.animatedboxing;

import android.app.Activity;
import android.content.Context;
import android.graphics.Canvas;
import android.graphics.Color;
import android.graphics.Paint;
import android.graphics.RectF;
import android.os.Bundle;
import android.os.Handler;
import android.os.SystemClock;
import android.view.View;

public class AnimatedBoxing extends Activity {
        private Handler handler;
        private PieView tPie; ❶
        private long pieEnds;
        private long fightLen=10*1000; // ms
        private long pieStarted; // round or break, ms

        @Override
        public void onCreate(Bundle savedInstanceState) {
                super.onCreate(savedInstanceState);

                handler = new Handler();
                handler.removeCallbacks(update);
                handler.postDelayed(update, 50);

                tPie = new PieView(this);
```

```
                    setContentView(tPie); ❷
        }

    private Runnable update = new Runnable() {
        public void run() {
            long now=SystemClock.uptimeMillis();
            if (pieEnds<now) {
                            pieStarted=now;
                            pieEnds=pieStarted+fightLen;
            }
            tPie.percent=
              (float)(now-pieStarted)/(pieEnds-pieStarted); ❸
            tPie.invalidate(); ❹

            handler.removeCallbacks(update);
            handler.postDelayed(this, 50); // ms ❺
        }
    };

    public class PieView extends View { ❻
        public float percent;
        public PieView(Context context) {
            super(context);
        }

        public void onDraw(Canvas canvas) {
            Paint paint = new Paint();
            paint.setColor(Color.WHITE); ❼
            RectF oval=new RectF(0, 0, ❽
                            canvas.getWidth(),
                            canvas.getHeight());
            canvas.drawArc(oval, 0, 360*percent, true, paint); ❾
        }
    };
}
```

Much of this code resembles previous examples, so let's focus on the animation. View.onDraw() draws a picture in different ways using different calls. Use
a timer to mark the view to be redrawn by calling the View.invalidate()
method approximately 20 times a second.

❶   The tPie view is stored as an attribute of the outer class, so you can
handle it from all of your methods.

❷   The animated custom view is made visible in the same way you made a
static picture visible in the previous example.

❸   Calculate the pie fill level as a decimal number so you know how large
a pie to draw. Just before the draw request, the timer's update.run()
calculates the fill level.

In the divisions of integers (int, long), the type must be changed to a
float. Otherwise, the result would be rounded to the closest integer.

The fill level of the pie is stored as a float in the variable percent. First,
the pie is completely empty, which corresponds to a fill level of 0, or 0%.
When the pie is totally full, the fill level is 1.0, or 100%. The percentage
value refers to one hundredth, which means that 1 equals exactly 100%.

The fill level is calculated by taking the time elapsed (`now-pieStarted`) and dividing it by the time spent for the whole pie (for a round or a break): `pieEnds-pieStarted`.

❹ Invalidate the view, which causes it to be redrawn (and hence, invokes the `PieView.onDraw()` method).

❺ The handler timer is responsible for invoking this object's (named update) `run()` method. Because you want smooth animations, the waiting period is short. The program uses 50ms (0.050 seconds) here, which makes the refresh rate (frame rate) 20 frames per second, or 20 hertz (1/[0.050s] = 20 1/s = 20Hz).

❻ A view must be defined as its own class that extends the `View` class. In contrast to the static image example shown previously, animations change between each `onDraw()` method call. In this case, the drawing is affected by variable percent, as you'll see in the `onDraw()` method.

❼ Most drawing commands use `Paint` class colors, which you can create immediately with names (`Color.GREEN`, `Color.RED`, and so forth) or as hexadecimal codes (`0xff22ffcc`). These hexadecimal codes start with `0xff`, because the first two digits are reserved for opacity (alpha channel).

❽ In Android, the `RectF` class defines a rectangle. The upper-left-corner coordinates are 0,0, and use the screen size for the lower-right-corner coordinates. Because you haven't set the program to use the full screen, the beams above push part of the pie beyond the bottom of the picture. You'll have a chance to fix this in a later example.

❾ The `drawArc()` method draws the pie (technically, a filled circular sector bounded by the specified arc) and takes several parameters:

— The size is specified with a rectangle (`Rect`) named `oval`. A fully filled pie would be the largest circle that fits this rectangle (i.e., the circle would touch all sides of the rectangle and would have exactly the same height and width as the rectangle).

— The fill level is specified by a beginning and end angle in degrees (`360*percent`). The degree of the angle of the pie's ending point changes continuously. A full circle is 360 degrees.

— The `drawArc()` method could also be used for drawing segments, in which case the second-to-last parameter (known as `useCenter`) would be `false` instead of `true` as shown here.

— The color is specified by the configuration of the `Paint` object named `paint`.

Figure 6-28 shows the result of running the program. Now you understand how to draw simple animations with Android.

Figure 6-28. *Animated pie*

Finishing the Clock

All the difficult parts are now complete, as you've tested them separately in small programs. You learned how to install the Android development environment and compile "Hello World." You applied some programming basics to create a graphical user interface. You built a timer. You drew to the cell phone screen using a custom view and created an animation by updating the custom view from a timer.

Now it's time to add the finishing touches to your app. Because few people check under the hood (or are qualified to do so), their opinion of your app is formed by its overall look and feel and usability, in the same way that you form your opinion of a product or service based on your customer experience. So let's put the rest of the components in place to make this app look polished.

Be warned: the final example code is long—over 200 lines. But the `import` lines alone are responsible for 10% of this. And more good news: most of the changes in this version of the code are small.

Put the resources in place

You can either create a new project, or simply use the previous example and replace the code in *AnimatedBoxing.java* with the example code shown later in this section.

As you did in the "Creating a Ringing Boxing Clock" project, create a new folder called *raw* under *res* in the Eclipse project. Next:

1. Copy the sounds, *fightsound.mp3* and *breaksound.mp3*, into *res/raw/*.

2. Copy the clock images, *clockbreak.jpg* and *clockfight.jpg*, to *res/drawable* (you will need to create the *drawable* directory).

3. Optionally, copy the program icon (*icon.png*) of three different sizes into *res/drawable-ldpi/*, *res/drawable-mdpi/*, and *res/drawable-hdpi/*, overwriting the files that are already in there. This will give you an attractive icon, as shown in Figure 6-29.

Pause the program when it's in the background

The program must end itself if the user moves to another program.

If you tested the earlier audio-based Boxing Clock examples thoroughly, you might have noticed a minor but irritating detail. The infernal ringing sound continues forever, even if you have navigated to another program by, for example, pressing the home button.

In addition to grating on your nerves, this also eats up your battery life, as do any unnecessary repetitive tasks that keep the processor from going into energy-saving mode. And updating the animations is useless if the user can't even see the picture on the screen.

Battery life is currently the main bottleneck for cell phone apps. It can even limit processing time, because a higher processing power eats up batteries quickly.

You can find all the image files in the example code at http://BotBook.com/.

Figure 6-29. *Boxing Clock icon in the Android app menu*

Chapter 6

Earlier, you made all the initializations in the activity's onCreate() method, which is invoked once in the beginning of the program. But it is not invoked when the user returns to the Boxing Clock from the home directory or from another program. The following example uses onPause() to handle this case properly.

Build the graphical user interface

We'll build a graphical user interface by stacking boxes on top of one another. The easiest of the Android containers is LinearLayout, shown in Figure 6-30.

Figure 6-30. *LinearLayout, stacking elements on top of one another*

Here's the new program code, using LinearLayout:

```
// bca08boxingClockReady - A usable boxing clock for Android.
// (c) Kimmo Karvinen & Tero Karvinen http://BotBook.com

package fi.sulautetut.android.animatedboxing;

import android.app.Activity;
import android.content.Context;
import android.graphics.Bitmap;
import android.graphics.BitmapFactory;
import android.graphics.Canvas;
import android.graphics.Color;
import android.graphics.Paint;
import android.graphics.RectF;
import android.media.MediaPlayer;
import android.os.Bundle;
import android.os.Handler;
import android.os.SystemClock;
import android.view.Gravity;
import android.view.Menu;
import android.view.MenuItem;
import android.view.View;
```

```
import android.view.Window;
import android.view.WindowManager;
import android.widget.LinearLayout;
import android.widget.TextView;
import android.widget.Toast;

public class AnimatedBoxing extends Activity {
    private Handler handler;
    private PieView tPie;
    private long pieEnds;
    private long fightLen=3*60*1000; // ms
    private long breakLen=60*1000;
    private long pieStarted; // round or break, ms
    private boolean fight=true; // will be changed immediately
    private MediaPlayer fightSound;
    private MediaPlayer breakSound;
    private static final int MENU105=105;
    private static final int MENU21=21;
    private static final int MENU31=31;
    private int boxingRedColor = 0xffff2704;
    private TextView tv;
    private long programStarted;
    private long roundsStarted=0;

    @Override
    public void onCreate(Bundle savedInstanceState) {
        super.onCreate(savedInstanceState);

        fightSound = MediaPlayer.create(this, R.raw.fightsound);
        breakSound = MediaPlayer.create(this, R.raw.breaksound);

        programStarted=SystemClock.uptimeMillis();

        LinearLayout container=new LinearLayout(this); ❶
        container.setOrientation(android.widget.LinearLayout.VERTICAL); ❷

        tv = new TextView(this);
        tv.setText("Welcome to BotBook.com Boxing Clock!");
        tv.setGravity(Gravity.CENTER); ❸
        tv.setTextColor(boxingRedColor); ❹
        tv.setBackgroundColor(Color.BLACK);
        tv.setPadding(5, 20, 5, 5); ❺
        container.addView(tv); ❻

        tPie = new PieView(this);
        container.addView(tPie);

        keepBackLightOn();
        fullscreen();

        setContentView(container); ❼
    }

    public void onPause() ❽
    {
        super.onPause();
        Toast.makeText(this, "Bye bye! "+statusMessage(),
Toast.LENGTH_LONG).show();
        handler.removeCallbacks(update);
    }
```

```
public void onResume()  ❾
{
    super.onResume();
    handler = new Handler();
    handler.removeCallbacks(update);
    handler.postDelayed(update, 50);
}

public void keepBackLightOn() {  ❿
    getWindow().setFlags(WindowManager.LayoutParams.FLAG_KEEP_SCREEN_ON,
                         WindowManager.LayoutParams.FLAG_KEEP_SCREEN_ON);
}

public void fullscreen()  ⓫
{
    requestWindowFeature(Window.FEATURE_NO_TITLE);
    getWindow().setFlags(WindowManager.LayoutParams.FLAG_FULLSCREEN,
                         WindowManager.LayoutParams.FLAG_FULLSCREEN);
}

public boolean onCreateOptionsMenu(Menu menu) {  ⓬
    menu.add(Menu.NONE, MENU105, 0, "10 s / 5 s");
    menu.add(Menu.NONE, MENU21, 0, "2 min / 1 min");
    menu.add(Menu.NONE, MENU31, 0, "3 min / 1 min");
    return true;
}

public boolean onOptionsItemSelected(MenuItem item) {  ⓭
    long now=SystemClock.uptimeMillis();
    pieEnds=now;
    programStarted=now;
    fight=true; // will be changed immediately because pieEnds=now
    roundsStarted=0;

    switch (item.getItemId()) {
    case MENU105:
        fightLen = 10*1000;
        breakLen = 5*1000;
        Toast.makeText(this, "10 second round, 5 second break. ",
                       Toast.LENGTH_LONG).show();
        return true; // will exit, so a "break;" would be
                     // unreachable
    case MENU21:
        fightLen = 2*60*1000;
        breakLen = 60*1000;
        Toast.makeText(this, "2 minute round, 1 minute break. ",
                       Toast.LENGTH_LONG).show();
        return true;
    case MENU31:
        fightLen = 3*60*1000;
        breakLen = 60*1000;
        Toast.makeText(this, "3 minute round, 1 minute break. ",
                       Toast.LENGTH_LONG).show();
        return true;
    }
    return false;
}

private String ms2mins(long ms)
```

```
    { // convert millisecond time to minutes and seconds
        int seconds = (int) (ms / 1000);
        int min = seconds / 60; ⓮
        int sec = seconds % 60; ⓯
        String pad;
        if (sec<10)
            pad="0"; ⓰
        else
            pad="";
        return ""+ min +":"+pad+sec;
    }

    private String statusMessage()
    {
        String s=""; ⓱
        long totalElapsed=SystemClock.uptimeMillis()-programStarted;
        s+=ms2mins(totalElapsed); ⓲
        s+=" elapsed. ";

        if (fight) {
            s+="Round ";
        } else {
            s+="Break ";
        }
        s+=roundsStarted+".";
        return s;
    }

    private Runnable update = new Runnable() {
        public void run() {
            long now=SystemClock.uptimeMillis();
            if (pieEnds<now) {
                fight=!fight;
                pieStarted=now;
                if (fight) { // round starts
                    roundsStarted++;
                    pieEnds=pieStarted+fightLen;
                    fightSound.start();
                } else { // break starts
                    pieEnds=pieStarted+breakLen;
                    breakSound.start();
                }
            }
            tv.setText(statusMessage());
            tPie.fightColors=fight;
            tPie.percent=(float)(now-pieStarted)/(float)(pieEnds-pieStarted);
            tPie.invalidate();

            handler.removeCallbacks(update);
            handler.postDelayed(this, 80); // ms
        }
    };

    public class PieView extends View {
        public float percent;
        public boolean fightColors=false;
        Bitmap fightBg;
        Bitmap breakBg;

        public PieView(Context context) {
```

```
        super(context);
        fightBg=BitmapFactory.decodeResource(getResources(),
                                       R.drawable.clockfight);
        breakBg=BitmapFactory.decodeResource(getResources(),
                                       R.drawable.clockbreak);
    }

    public void onDraw(Canvas canvas) {
        canvas.drawColor(Color.BLACK);

        Paint paint = new Paint();
        paint.setColor(Color.WHITE);

        float w=canvas.getWidth(); ⓳
        float h=canvas.getHeight();
        float margin=10; // pixels from pie to nearest wall
        float r=(Math.min(h,w)-margin)/2; // ray, half diameter ⓴
        // center first - because we want it centered ㉑
        float cx=w/2;
        float liftCenter=-50;
        float cy=h/2+liftCenter;
        // top left ㉒
        float tx=cx-r;
        float ty=cy-r;
        // bottom right ㉓
        float bx=tx+2*r;
        float by=ty+2*r;

        RectF bgRect = new RectF(tx, ty, bx, by); ㉔

        RectF oval=new RectF(bgRect); // copy bgRect measures instead ㉕
                                  // of creating a reference
        oval.inset((int)(r*0.29), (int)(r*0.29)); ㉖

        if (fightColors) { ㉗
            canvas.drawBitmap(fightBg, null, bgRect, null);
        } else {
            canvas.drawBitmap(breakBg, null, bgRect, null);
        }

        canvas.drawArc(oval, -90, 360*percent, true, paint); ㉘
    }
};
}
```

Here's what's going on in the code:

❶ Create the LinearLayout and add widgets (buttons, views) to it from the top. Name it container, because the layout functions as a container.

❷ Configure the layout to stack items vertically.

❸ Customize the text view, configuring it to center items.

❹ Choose the coloring of the letters with setTextColor() and the background with setBackgroundColor().

❺ Add some padding to the text view, so it's not flush against its containers. The parameters are (left, top, right, bottom). If you don't want to memorize them, you can always see the parameters required in the Eclipse by hovering your cursor over the method name in the code editor.

❻ Create the views one at a time and place them into the container.

❼ Make the layout visible with `setContentView()`.

❽ Handle an early departure from the program by overriding the `onPause()` and `onResume()` methods.

When a user navigates away from the Boxing Clock (for example, by pressing the home button), `onPause()` runs automatically. When this happens, a text message such as "Bye bye! 12:03 elapsed. Round 4" pops up. This section also disables the timer that would otherwise keep calling our main loop, halting sounds and animations.

❾ When the user returns to the program, `onResume()` runs, starting animations and sounds again.

The code in the `onResume()` method was moved from the old `onCreate()` method used in earlier examples.

The only change from that method is to now call `super.onResume()` (the `onResume()` method of the parent class).

❿ During a boxing match, it's inconvenient to have to press the cell phone buttons constantly to keep the screen from dimming, so switch the background light on whenever the Boxing Clock is visible. The background light uses batteries heavily, so take care.

⓫ You can set the window to a full-screen mode by hiding additional embellishments. It is a matter of taste whether the program looks better in full screen, and you can remove this function call if you want to see the top bars (`keepBackLightOn()` and `fullscreen()` are both called from `onCreate()`).

⓬ Within Android, you can make a menu visible by pressing the cell phone's Menu button, which is usually a physical button on the cell phone. This section makes a menu for the program. It uses a number (such as `MENU105`, defined toward the top of the class) to make it recognizable as the same choice within different methods. Then, it creates a menu by bypassing the `Activity.onCreateOptionsMenu()`.

Any integer will suffice as the menu identification number. These numbers are normally stored in constants that are global to the entire class. In Java, `private static final` refers to a constant, which is similar to `const` in some other languages.

Menus are created in the beginning of the program. If you have many choices, you divide them into groups. Here, the groups are unnecessary, so all choices are in the group `Menu.NONE`. Android recognizes the menu based on a number that is marked here with the constant `MENU105`. The order of our choices doesn't really matter, so give them all the order number `0`. Give the menu some short text, visible to the user, such as "10 s / 5 s" (10-second rounds, 5-second breaks).

⓭ When the user makes a choice from the menu, call `onOptionsItem Selected()`. The number of the choice is passed as a parameter of this function, and the numbers are the same as those specified when the menu was created. The `case` statement takes the appropriate action based on which menu was selected.

⓮ In `ms2mins()`, convert milliseconds to strings displaying minutes and seconds.

⓯ Because we are dividing integers, the decimal part is automatically cut off. For example, when performing integer division, `1/2 = 0`.

The truncated seconds can be determined from the remainder, which is calculated by the modulus operator: `%`.

⓰ If the seconds part of the time has only one digit, pad it with a leading zero (so 9 becomes 09).

⓱ In the `statusMessage()` method, create an empty text string, s, in which to add text word by word with the compound addition operator `+=`.

The compound addition operator is just a shorter way of writing "the string takes on its original value plus the new string as its value." These mean the same thing:

```
s += ", A";
s = s + ", A";
```

⓲ Users want to see minutes and seconds, but this program handles time in milliseconds. Call `ms2mins()` to reformat the time for the user.

⓳ Cell phone displays vary in size and shape. You can even rotate some cell phones to change the display orientation. Therefore, it is good practice to place widgets and images in relatively defined positions.

Begin by giving some handy names (h, w) to the canvas height and width. Set the margin here, too.

⓴ Set the pie's radius so that it runs to the edge of the screen.

㉑ The pie goes in the middle, so calculate its center point (cx, cy; short for center y and center x).

㉒ To position a circle, we define a rectangle in which the circle fits exactly. So the circle touches the `Rect` at the midpoint of each side. Pies work just like circles, as a full pie is a circle.

We already know the center point of the circle (cx, cy). The bounding `Rect` has the same center. The walls are one radius r away from the center. Thus, we can get the coordinates of the bounding box with simple addition and subtraction. Figure 6-31 illustrates this.

㉓ The bottom of the bounding box is one diameter (2r) from the top of the box. The right wall is one diameter away from the left wall.

㉔ Create a `RectF` class square object, because the method used for drawing the pie requires a `RectF` as its parameter. The background image will be placed using this object (bgRect).

The pie must be slightly smaller than the screen for the clock's edges to remain visible. The center point automatically hits the correct spot, because the center point of the background image is exactly in the middle of the 400×400-pixel image.

㉕ To make two different size rectangles, make a copy of the rectangle (oval) and modify it in the next line.

㉖ Reduce the square framing the pie by 29%.

㉗ Earlier, the PieView constructor loaded a visible background onto a variable during a round. You can now draw it on the screen.

The parameters for Canvas.drawBitmap() are an image (fightBg) and its placement (bgRect). Assuming you don't have a preference for the color (the last parameter) and you don't want to clip the image (the second parameter), give these parameters a value of null.

㉘ Draw the pie, framed within the rectangle you calculated. Turn the starting angle 90 degrees counterclockwise so that the pie will start on the top of the screen. The fill level is a percentage of the 360 degrees of the whole turn. You want a pie (a circular sector), not an arc, so specify true as the useCenter parameter. The color of the pie is paint, which was defined earlier.

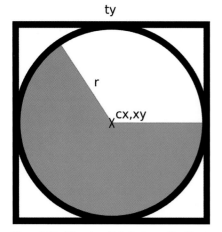

Figure 6-31. *The pie within its bounding Rect*

The Boxing Clock is now ready to use (see Figures 6-32 and 6-33).

Figure 6-32. *Finished Boxing Clock, including sounds, graphics, and menus*

Figure 6-33. *Choosing the length of a round*

Before you go off to start practicing, let's review what you have learned here. First, you installed the Android development environment. Testing programs in the emulator is quick. You can now create simple animations within the main loop. You can play sounds from MP3 files. And most important of all, you are familiar with Android's way of constructing programs.

Installing on the Physical Phone

It's hard to drag a computer to the gym, so let's install our program on the cell phone.

Enable USB debugging

First, you need to configure your phone to support debugging over USB. Go to the home view by pressing the cell phone's home button. Next, tap the Menu button and choose Settings→Applications→Development. Turn on "USB debugging." If you want to keep the phone from switching off while you are testing apps, you can set the backlight to stay on whenever the phone is plugged in via USB (turn on "Stay awake"). Figure 6-34 shows the steps needed to configure these settings.

You will find it helpful to enable the "Stay awake" option while you are debugging. If you don't enable this, you may need to manually wake your device up before you try to run a program on it.

Figure 6-34. *Switching on the "USB debugging" and "Stay awake" options from the phone*

Make the physical connection

Next, connect your phone to your computer via a USB cable. Return to Eclipse.

Now you need to choose the DDMS *perspective* from Eclipse. Perspectives control which windows are shown within Eclipse. You can think of a perspective as a task-oriented Eclipse window configuration. The DDMS perspective is for device debugging.

Choose Window→Open Perspective→Other→DDMS. The Eclipse view changes to the view shown in Figure 6-35.

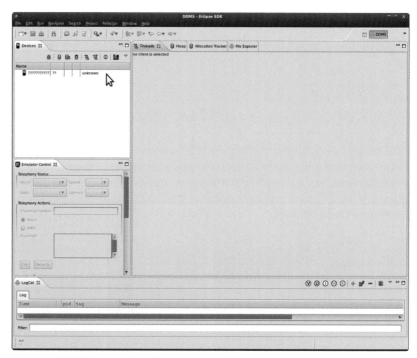

Figure 6-35. *DDMS perspective showing an unconfigured device*

Where is the phone, then? If things worked out, you'll see an entry for your phone at the upper left of the screen. On Linux or Windows, you might need to perform some additional configuration, as described next.

Configure Linux for DDMS

If you see question marks (??????????) and "unknown" devices on the upper-left side of the Devices tab next to a phone icon, you'll need to make some changes.

Close Eclipse. Open a Terminal window, change directory to your home directory (or wherever the Android SDK was installed), and list the devices known by Android:

```
$ cd
$ ./android-sdk-linux_*/platform-tools/adb devices
List of devices attached
???????????? no permissions
```

You'll need to configure Ubuntu to give normal (nonroot) users access to the device. First, you'll need to determine the USB vendor ID for your phone. Issue the lsusb command (you might need to provide your password when prompted):

```
$ sudo lsusb
[sudo] password for user: *********
...
Bus 001 Device 003: ID 18d1:4e12 Google Inc. Nexus One Phone (Debug)
```

This displays a long list of devices, but you can recognize your phone by its name. The list also shows the VendorID (the first four numbers of the ID). Here, the VendorID is 18d1. To give all users permissions for this device, you need to create a new *udev* rule file. Issue this command to open the new file in a text editor (the `sudo` command is necessary to edit the file with root/admin privileges):

```
$ sudo nano /etc/udev/rules.d/99-android.rules
```

Type the following into the file, replacing `18d1` with the correct VendorID:

```
## For using physical Android phones in DDMS and Eclipse
SUBSYSTEM=="usb", SYSFS{idVendor}=="18d1", MODE="0666"
```

Save the file by pressing Ctrl-O and then pressing Enter or Return. Press Ctrl-X to exit the text editor. Then restart the *udev* system:

```
$ sudo service udev restart
```

Unplug and reattach the USB cable to the phone. You should now see the device name when you run the `adb devices` command, and the phone will show up in the Devices tab of the Eclipse DDMS perspective (Window→Open Perspective→Other→DDMS), as shown in Figure 6-36.

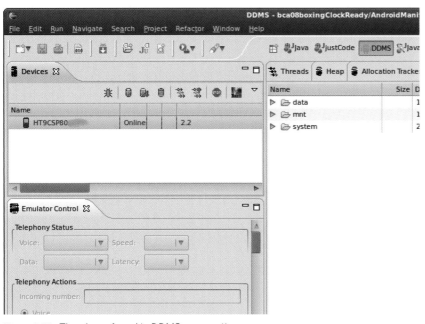

Figure 6-36. *The phone found in DDMS perspective*

Configure Windows USB settings and drivers

To install the USB driver:

1. Make sure you're logged in as a user with administrative privileges. (Remember to log back in as a normal user after you are done with configuration.)

2. From within Eclipse, click Window→Android SDK and AVD Manager.

3. Click Available Packages and expand the list.

4. Choose the USB Driver Package and click Install Selected (Figure 6-37). Then follow the prompts to install the driver.

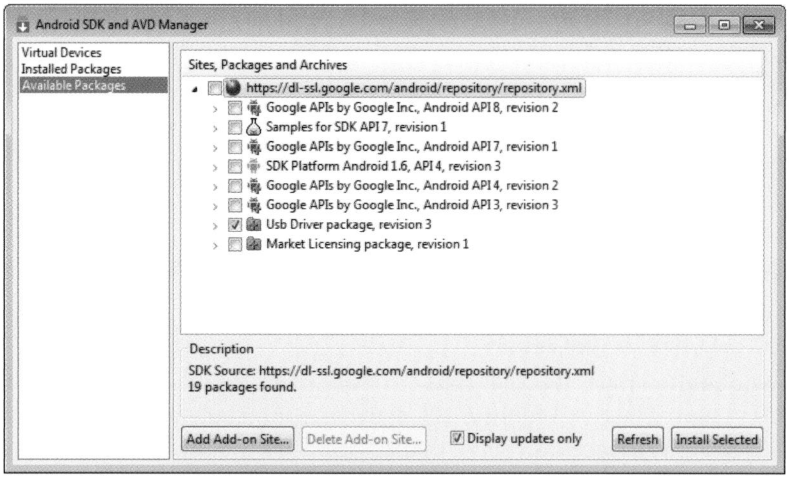

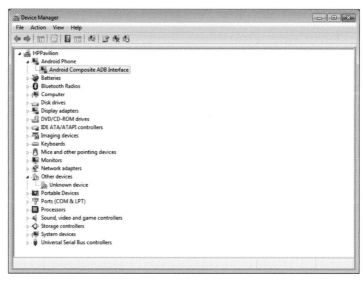

Figure 6-37. *USB driver download package*

If you're using something other than a Nexus One phone, you should still be able to use the driver, but check http://developer.android.com/sdk/win-usb .html for more details. As of this writing, Windows 7 is not yet supported, although Google's own phone works with the preceding instructions. The driver and instructions can be found for XP and Vista.

Connect your Nexus One via a USB cable. Open the Start menu, right-click Computer, and choose Properties. System Properties will open. Select Device Manager from the left.

Under Other Devices, open Android Phone, right-click the Composite device underneath it, and select Update Driver Software. Browse your computer for the driver software, choose *C:\android-sdk-windows\usb_driver* (the actual path depends on where you installed the SDK), click Next, and install the driver. Figure 6-38 shows a successful installation.

Figure 6-38. *The Nexus correctly installed*

Run the app on the phone

Open Eclipse and move to the DDMS perspective (Window→Open Perspective→Other→DDMS). The phone is now visible in the Devices tab with its own ID number. Next to that you should see the text "Online" (Figure 6-39).

Figure 6-39. *The phone appearing in Eclipse*

Return to the Java perspective (Window→Open Perspective→Java). Click the Run button on the toolbar. Now Eclipse will ask whether you want to run its emulator in a virtual machine or in the real phone (Figure 6-40). Try out the phone (Figure 6-41)!

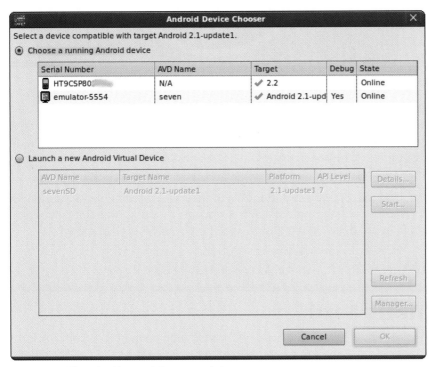

Figure 6-40. *Choosing the emulator or a real phone*

Figure 6-41. *Program starting on a real phone*

Differences Between the Emulator and Phone

The emulator is a handy tool when you're programming, but it doesn't always replicate the phone in all respects. Therefore, when moving to a real phone, you should expect some surprises.

For example, the emulator was not rotated while we were testing, but people frequently rotate their phones. When a phone is turned sideways, Android automatically turns the view into a horizontal (landscape) mode. However, we didn't make any accommodation for this in our code, so the rotated view does not look good (as shown in Figure 6-42). Our options are to make the graphical user interface support multiple modes or forbid the turning of the images. We chose the latter, adding this line to the onCreate() method to keep the image in portrait orientation:

```
setRequestedOrientation(ActivityInfo.SCREEN_ORIENTATION_PORTRAIT);
```

Figure 6-42. *A cell phone turned to its side messes up the image's appearance*

We developed the program for the most popular Android version: 2.1 API level 7. When we checked one of the phones we tried to run it on, we noticed it supported only 1.5 API level 3, so we installed the support for that SDK. In Eclipse, we chose Window→Android SDK and AVD Manager and installed support for Android 1.5 API 3 in the same way you installed 2.1 earlier.

We changed our project settings file to the lowest acceptable API version (3) by double-clicking AndroidManifest.xml and choosing AndroidManifest.xml from the tab below its window. We set the minSdkVersion as shown:

```
<uses-sdk android:minSdkVersion="3" />
```

When we tried to run the project again, holy smokes, it opened on the phone!

What's Next?

You have now learned the basics of Android programming. You know the program structure and can create animations and play music. You will be able to install your program on a real phone. Reward yourself with some boxing, or just jump rope. Have fun!

Be brave when developing your ideas! A cell phone has many advantages over a computer. A cell phone is always accessible. It has many sensors already built in, such as a video camera, microphone, and accelerometer. Cell phones can be used for connecting to computers over a wireless network. Which programs have you been missing on the road?

Remote for a Smart Home

In this project, you will build a computer-based remote to control AC-powered devices. Using the techniques you'll learn here, you can go on to develop control systems all over the home, such as cell phone–controlled lighting or curtains that open or close on a timer.

You'll begin this project with *relays*, electromechanical switches you can use to replace manual buttons with ones controlled from the Arduino. Next, you'll disassemble a standard remote control (meant for household functions such as lights) to use as a platform for an Arduino-based remote. Using a remote is a safe approach, because you don't have to touch the AC (alternating current) side of the circuit. You'll solder terminals for relay control to the remote's circuit board.

To enable the Arduino to receive instructions from a computer, you'll first need to send it instructions over the serial port. Then, you'll connect the remote to both the Arduino and the relays, after which you'll control the larger circuit from a computer. Finally, you'll create a graphical user interface within Python to control the whole package.

Before starting this project, you must install the PyGTK development environment and test the "Hello World" code covered in the Interactive Painting project in Chapter 5. It is also helpful to practice soldering a bit beforehand, because now you will have to solder wires straight to the circuit board instead of simply joining them together. You can practice soldering by building the Stalker Guard project in Chapter 3.

Figure 7-1 shows what the finished project looks like.

The PowerSwitch Tail (http://www. makershed.com/ProductDetails. asp?ProductCode=MKPS01) is a handy device that lets you switch AC voltages without endangering your Arduino. It's not wireless, though, and you won't have the fun of taking apart a remote control and modding it.

What You'll Learn

In this chapter, you'll learn:

- How to control AC-powered devices

Figure 7-1. *The finished project in an enclosure*

- How to use relays
- The basics of creating graphical user interfaces

Tools and Parts

For this project, you'll need the following tools and parts, which are shown in Figure 7-2.

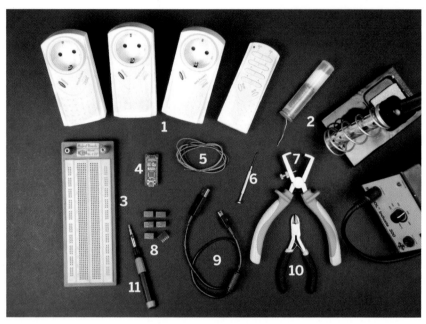

Figure 7-2. *Tools and parts needed for this chapter*

1. Remote control–equipped AC sockets (such as the Stanley 31164 Indoor Wireless Remote Control with Single Transmitter), available online or at home improvement stores.

2. Soldering iron and solder.

3. Solderless breadboard (SHED: MKEL3; EL14: 15R8319; SFE: PRT-00112).

4. Arduino Nano (SHED: MKGR1; *http://store.gravitech.us;* or *http://store.gravitech.us/distributors.html*).

5. Jumper wire in three colors (SHED: MKEL1; EL14: 10R0134; SFE: PRT-00124).

6. Small flat-head screwdriver.

7. Wire strippers (EL14: 61M0803; SFE: TOL-08696).

8. Six 5V sensitive relays (EL14: 64K3159).

9. USB cable.

10. Diagonal cutter pliers (EL14: 52F9064; SFE: TOL-08794).

11. Phillips (cross-head) screwdriver.

The Relay: A Controllable Switch

A relay connects two pins to each other when current is applied across its control pins. The two sides of the relay are isolated from each other (for example, in most mechanical relays, the control pins activate an electromagnet that causes the switch to move).

Because the control pins and the switch are electrically separated, a relay can be used to control large current with a small current. You can also use it to replace an external device's button, which allows you to control the device from the Arduino while keeping the Arduino isolated from it.

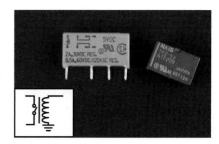

Figure 7-3. *Two 5V relays*

This chapter uses a 5V relay, two examples of which are shown in Figure 7-3. The Arduino digital pins use 5 volts, so it's enough to trigger the relay.

The Arduino controls the relay in Figure 7-4. Connect digital pin 2 on the Arduino to the relay pin marked D2, and the Arduino ground to the relay pin marked GND. Connect the external component you wish to control to the X1 and X2 pins. When D2 is switched on (HIGH), the relay will connect pins X1 and X2 to each other.

Toggling a Relay with Arduino

Figure 7-5 shows the Arduino Nano seated in a breadboard along with a relay and an LED. Figure 7-6 shows the circuit diagram.

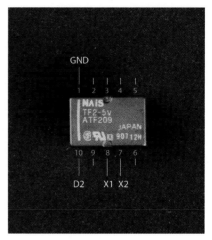

Figure 7-4. *Relay circuit connections*

> *The LED is powered by the Arduino, even though relays are typically used to electrically isolate components from one another. By the end of the chapter, we'll be using relays as they were meant to be used, with the device to be switched (the wireless remote control) being fully isolated from the Arduino.*

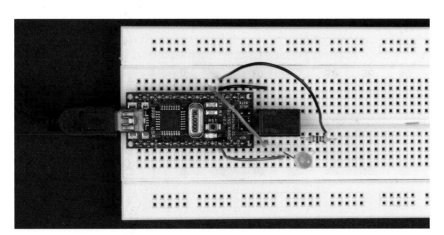

Figure 7-5. *Connecting the LED with a relay*

Here's how to set up the circuit:

1. If you are using a Nano, Boarduino, Bare Bones Board, or similar, insert it into the prototyping breadboard.

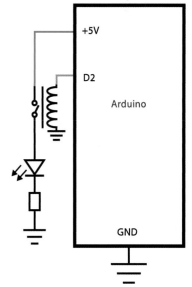

Figure 7-6. *Circuit diagram with Arduino, relay, LED, and resistor*

2. Connect the Arduino to the relay's control pins:

   a. Insert the relay into the prototyping breadboard so that the relay's white stripe is adjacent to the Arduino. This situates pin 1 of the relay in the upper-left corner. If your relay does not have a stripe, consult its datasheet (usually available online from the vendor where you purchased it) to determine the pin orientation.

   b. Connect the Arduino digital pin 2 to relay pin 10 (bottom left) with a green jumper wire.

   c. Connect the Arduino ground (GND) to relay pin 1 with a black jumper wire.

3. Connect the circuit to be switched:

   a. Connect Arduino's +5V pin with a red jumper wire to relay pin 8, the middle one in the bottom row. Pin 8 is marked as X1 in Figure 7-4.

   b. Connect the LED's positive (longer) leg to relay pin 7. Pin 7 is in the bottom row, second from the right. It is marked X2 in Figure 7-4.

   c. When the Arduino activates the control pins (by taking D2 high, +5V), X1 and X2 will be connected, lighting the LED.

4. Connect the LED to ground:

   a. Connect the LED's negative (shorter) leg to a resistor. To do this, insert the LED's negative leg to the prototyping breadboard's hole in one of the free vertical rows, and then place one end of the resistor in the same row (and on the same side of the breadboard) as the LED.

   b. Connect the other end of the resistor to the Arduino GND pin. If necessary, use black jumper wire, as shown in Figure 7-5.

The following program is similar to the Blink program in Chapter 2. The code toggles the relay on for one second and then off for one second. An LED connected to the relay shows you whether it's on or off. The relay clicks quietly when it's on; you can easily hear this by placing your ear close to it.

```
// ledRelay.pde - Control led with a relay
// (c) Kimmo Karvinen & Tero Karvinen http://BotBook.com

int relayPin = 2;
void setup()
{
  pinMode(relayPin, OUTPUT);
}

void loop()
{
  digitalWrite(relayPin, HIGH);
  delay(1000);
  digitalWrite(relayPin, LOW);
  delay(1000);
}
```

If you're using a different relay, replace pins 10 and 1 with whichever two pins activate the relay's coil, and pins 7 and 8 with whichever two pins are the relay switch's contacts.

You can generally find this information in the relay's datasheet.

If you are using another model of Arduino, you can set up the relay and LED in the breadboard and connect the Arduino with jumper wires.

As you can see, the code functions just like the regular LED blinking code, but the `ledPin` variable is changed to a `relayPin` and the pin number is changed to 2, making the relay switch the LED on and off for a second.

Hacking the Remote Control

By using a relay, you can control the buttons of almost any kind of device. As the basis of this smart home remote, we used an AC socket remote control that we bought from Clas Ohlson in Helsinki (see Figures 7-7 and 7-8). You can find many similar products to use, such as the Stanley Indoor Wireless Remote Control and the Woods Indoor Wireless Remote Control Outlet.

AC current is dangerous. That is why we don't want to touch the actual outlet, only modify the low-current remote. Don't play with high current if you don't have sufficient knowledge and training. Laws and regulations controlling radios and electric power vary by country. You are responsible for obeying your local laws.

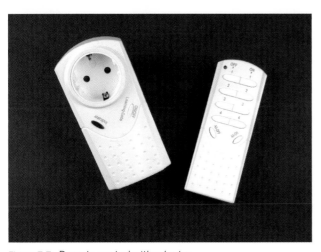

Figure 7-7. *Remote control with adapter*

Figure 7-8. *Back of remote control*

Before doing anything else, make sure that the remote control functions. Connect the three receivers (remote-controlled power outlets) to the wall AC socket and plug in something (such as a lamp). Test that you can switch every device on and off easily. With a full battery, the receivers should react in less than a second.

You're going to be taking apart the receivers. Reassembling them is easier if you can find all the parts. Place the parts into a cup or an egg carton, rather than scattering them all around the table.

Disassembly

First, remove two Phillips-head screws from the bottom of the remote. Now the enclosure is attached only with plastic clips.

Before you disassemble the remote, you might want to mark the positive lead of the battery terminal to make it easier to place the battery the right way. Do not mark it with anything that will block the voltage of the battery. If your remote uses a 9V battery, this step will not be necessary. Some devices have unusual batteries, so it is useful to write its model number down as well.

Some cases have a hole, which may be covered by tape. This hole is typically used to access the jumpers that select which radio channel the remote control should broadcast on. You can always make such a hole with a Dremel tool. If you'd rather reassemble the case and continue to use the remote's buttons, you can run the wires through this hole after you've soldered them into place.

Open the battery compartment and remove the battery. Near the battery compartment, in the seam between the top lid and the base of the remote, you'll find a small gap. Insert a flat-head screwdriver into the gap, as shown in Figure 7-9, and pry the two halves apart. A light twist is enough. Figure 7-10 shows the opened remote controller.

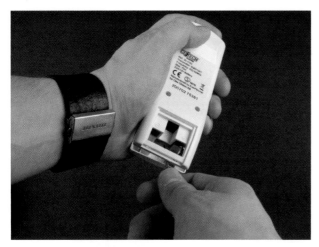

Figure 7-9. *Opening with a screwdriver*

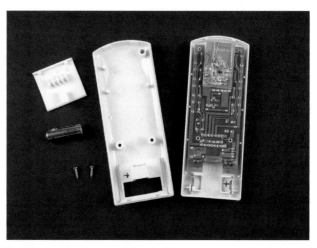

Figure 7-10. *Opened remote control*

Remove the circuit board by lifting it lightly with the screwdriver, as shown in Figure 7-11. The remote control is now disassembled, and the detached circuit board is shown in Figure 7-12. The metallic battery holder and the radio section of the circuit board are sturdily attached; do not remove them. The radio section is a small, separate daughterboard simply connected with solder to the front of the larger circuit board. It can break relatively easily, so do not twist or bend it.

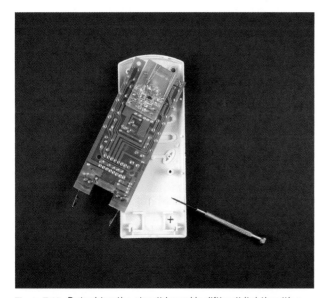

Figure 7-11. *Detaching the circuit board by lifting it lightly with a screwdriver*

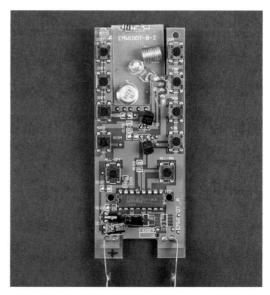

Figure 7-12. *The circuit board*

Chapter 7

Testing

Test that the remote control still works when disassembled. To do so, connect the battery, press the remote control buttons, and make sure that the remote outlet turns on and off. Now you know that you have not destroyed the device while taking it apart, that the battery has power, and that the controlled switch functions.

It is easiest to test the opened remote control by placing it on top of the case's base. The battery will stay in place when the battery compartment lid is closed. For observation and soldering, lift the circuit board apart from the base.

You can also test the functions of the remote control's buttons by connecting the solder pads of the switches with a length of jumper wire; this is exactly what you'll use the relay to do.

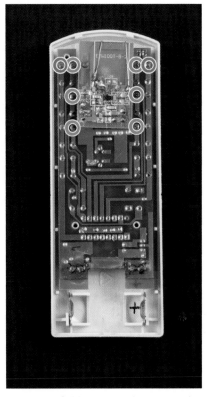

Figure 7-13. *Solder pads: red wires to red ones and green wires to green ones*

Soldering

For each button, cut one piece of green jumper wire. Depending on the configuration of your remote, you will need either one or two pieces of red jumper wire. You'll use the green wire for each button and the red wire for the common connection that each group of buttons goes to.

Make each wire at least 15cm long. Solder the wires to the circuit board on the solder pads marked in Figure 7-13.

Your remote control will almost certainly differ, so you might need to use a multimeter and simple observation to determine which solder pads correspond to which button. These pads control the remote-controlled power outlets and can switch them on or off.

Figure 7-14 shows the Stanley remote control, which has a different configuration. The pad marked in red is the single common connection for all the buttons. The pads marked in blue are the pads corresponding to each button. We found the pads to be extremely fragile, and ended up damaging one of them.

However, in the end we found a less fragile location to connect our wires: the four diodes visible in Figure 7-14 to the right of the top two pads, and the two diodes just above the bottom group of pads. These diodes had ample amounts of solder already on them, and were much easier to solder wire to (we soldered to the pad opposite the stripe on each diode).

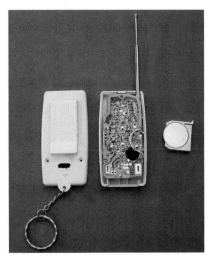

Figure 7-14. *Solder pads on the Stanley wireless remote control*

Heat the solder joint for one or two seconds, and touch the solder to the joint (it should melt in about one second; if not, your iron is not hot enough). Within one second, pull the solder away and then immediately remove the soldering iron from the joint. Figure 7-15 shows us soldering the wires to the board, and Figure 7-16 shows the finished job.

> *It is a bit more difficult to solder to a circuit board that already has solder joints on it. You can use a liquid flux pen (such as the one at* http://www.sparkfun.com/products/8967) *to make your job much easier.*

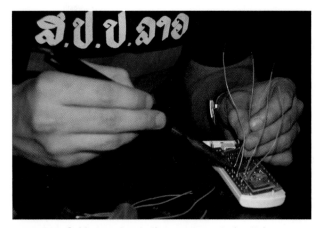

Figure 7-15. *Soldering wires to the remote control switches*

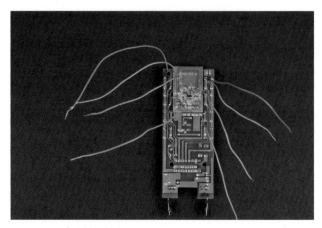

Figure 7-16. *Soldered wires*

Controlling the Arduino from the Computer

Because you want to control the remote from the computer, you need a way to communicate between the devices. The serial port is the most common way to connect Arduino to more complex devices, such as computers, cell phones, or RFID readers. This example uses serial ports via USB, but you can also use serial ports wirelessly—for example, over Bluetooth.

Toggling an LED with the Serial Port

Try the following program by giving commands from the Arduino development environment serial console. By sending either 8 or 2, you can switch the internal Arduino LED on and off, respectively.

```
// ledSerialControl.pde - Control led via serial
// (c) Kimmo Karvinen & Tero Karvinen http://BotBook.com

int relayPin=13; ❶

void setup()
{
    pinMode(relayPin, OUTPUT); ❷
    Serial.begin(9600); ❸
    Serial.println("Control relay with serial - ready. www.BotBook.com");
}

void loop()
{
    char ch = Serial.read(); ❹
    if (ch=='2') { ❺
        digitalWrite(relayPin, LOW); ❻
    }
    if (ch=='8') { ❼
        digitalWrite(relayPin, HIGH);
    }
}
```

Let's take a look at the code:

❶ First, declare a global variable. Because the variable has been declared outside of all functions, it is global and can therefore be used in all functions. So, you can refer to the `relayPin` variable functions `setup()`, `loop()`, and all other functions.

❷ Setting the pin you're using to the `OUTPUT` state allows you to later set it using the `digitalWrite` function. This way, for example, you can light up and switch off an LED hooked up to it, or switch a relay on or off.

❸ Open the serial port. As the serial port opens, its data transfer rate is also specified, usually at the default speed of 9,600 bits per second (bps) used by the Arduino development environment. This speed is enough to transfer short messages and numbers. The data rate 9,600bps equals approximately 1kB/s (kilobytes per second), the data rate in the application layer. Compared to the transfer rate for web browsing on a computer, data moves relatively slowly over the serial port.

With the serial port in use, you can print data about the state of the program to the serial console. A short welcome message indicates that the program's `setup()` function is ready.

❹ `Serial.read()` returns one character read from the serial port. One character (`char`)—for example, `'a'`, `'X'`, `'#'`, or `'6'`—is 1 byte (B) long, which equals 8 bits. If there are no characters in the queue to be read in the serial port, `Serial.read()` will wait. The program execution pauses during the wait. When a character arrives from the serial port, the execution of the program continues. A character is stored in a new `char` type variable as defined here, named `ch`.

❺ Check whether the character in the `ch` variable is the command you're looking for. If the comparison `ch=='2'` is `true`, which means the `ch` value is the character `'2'`, the block after the `if` statement executes. The C language uses two equals signs (`==`) to compare for equality.

> *Character `'2'` is not number 2. The character has single quotes around it; the integer 2 does not. ASCII character values are stored as numbers. For example, `'a'==97`; `'2'==50`; and `'8'==56`. So the numeric value of the byte is not the same as the character it represents. For more information, search the Web for "man ascii" or "ascii chart."*

❻ When this condition is met, the program sets the `relayPin` pin to `LOW` state, which means 0 volts (off), and switches off the LED connected to the pin. If the pin has a relay, it switches off the connection between the pins that it controls.

❼ In a similar way, the program switches on the pin if the value of the `ch` variable is the character `'8'`.

When the program reaches the end of the `loop()` function, it runs it again until the Arduino is switched off. After you send the sketch to the Arduino, click the Serial Monitor icon in the toolbar (Figure 7-17); then type **8** or **2** and click Send.

Figure 7-17. *Serial Monitor button*

Change the following line of code:

```
int relayPin=13;
```

to this, which is the first step toward controlling a relay from the serial port:

```
int relayPin=2;
```

Hook up the Arduino to the circuit shown earlier in "Toggling a Relay with Arduino." Next, upload the sketch to the Arduino, open the Serial Monitor, and send an 8 or 2 to the Arduino to control the relay.

Connecting Relays to the Switches

The following circuit is similar to the previous circuit, which blinks an LED by means of a relay, but this example uses a remote control button instead of an LED and a resistor. Here's how to connect the first button:

1. Insert the Arduino into the prototyping breadboard. Insert the relay into the board so that the white stripe is on the left, to the right of the Arduino, as shown in Figure 7-18. If your relay does not have a stripe, consult its datasheet (usually available online from the vendor where you purchased it) to determine the pin orientation.

2. Connect the relay control lines. To do so, connect the Arduino GND with a black wire to relay pin 1 (above the white line). Connect the Arduino D2 pin with a green cable to relay pin 10 (under the white line).

3. Now, connect the device you want to control to the relay. Locate the off button for the remote control's first outlet, and connect that button's wires to relay pins 7 and 8 (the second from the right and the middle, respectively, in the bottom row). The relay is just a controllable switch and is not polarized, so you can connect the wires either way. Figure 7-19 shows the circuit diagram.

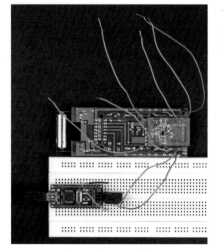

Figure 7-18. *One relay controlling one button*

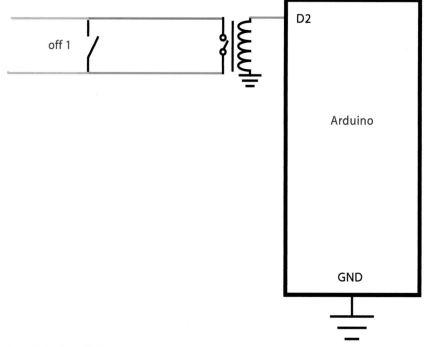

Figure 7-19. *Circuit diagram*

Chapter 7

The remote control has three controllable receivers, with two buttons (on and off) each. In all, six buttons require six relays. Wire up the remaining buttons, connecting one relay to each button. One pin controls each relay.

The circuit follows the same pattern as the first switch you wired up:

1. Green wires from the remote connect to each relay pin 7.

2. Red wires from the remote connect to each relay pin 8. Because each side of the remote shares a common red wire, you can use jumper wire (see the upcoming note) to connect each group of three relays to one red wire.

3. The Arduino digital pins (see Table 7-1) connect to each relay pin 10.

4. The Arduino ground connects to each relay pin 1. Because there is a common GND wire from the Arduino to the relay (pin 1), you can connect the Arduino to the prototyping breadboard's top horizontal ground row and connect the relays to it with short jumper wires, as shown in Figure 7-20. Note that we've moved the connections to the remote control from the bottom half of the board—as shown back in Figure 7-18—to the top; this is possible because the relay has another set of pins across from the ones we used back in Figure 7-18. Figure 7-21 shows the circuit diagram.

> *Note the black jumper in the top middle of Figure 7-20. On most full-size breadboards, the top and bottom horizontal rows are continuous across them all. If your prototyping board has a gap like the breadboard shown in the figure, you might need to add a jumper as we did here.*
>
> *Also note that there is no such red jumper. This is because the remote control we used has two common lines on each side of the remote, and we needed to keep them separate. If your breadboard does not have this gap, you should not use the horizontal row for the red wires. Instead, connect the common line to the first relay in each group of three and add one jumper wire each for the remaining two relays in each group.*
>
> *If your remote control has only one common line, as is the case with the Stanley remote, you don't have to worry about separating your common lines.*

Table 7-1. *Arduino pins to remote buttons*

Arduino pin	Remote button
Digital 2	Outlet 1 off
Digital 3	Outlet 2 off
Digital 4	Outlet 3 off
Digital 8	Outlet 1 on
Digital 7	Outlet 2 on
Digital 6	Outlet 3 on

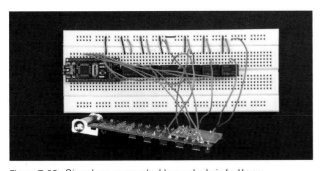

Figure 7-20. *Six relays connected to control six buttons*

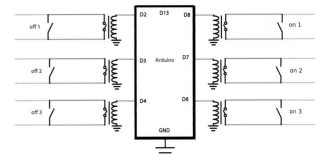

Figure 7-21. *Circuit diagram*

Six-Switch Code

How can you update your code to handle multiple pins? One option is to create a whole bunch of `if` statements for every possible port number, but then the code would be too long and have a lot of repeated segments. We solved the problem by encoding the digital pin number in the message we send to the Arduino.

The following code "pulses" the specified digital pin (0–9) by taking it high for one second and then low again, which is the equivalent of pressing the button for one second.

```
// sixRelays.pde - Control relays with a computer
// (c) Kimmo Karvinen & Tero Karvinen http://BotBook.com

ledPin=13; ❶

void relayPulse(int pin) ❷
{
    pinMode(pin, OUTPUT); ❸
    digitalWrite(ledPin, HIGH); ❹
    digitalWrite(pin, HIGH);
    delay(1000); ❺
    digitalWrite(pin, LOW); ❻
    digitalWrite(ledPin, LOW);
}

int charToInt(char ch) ❼
{
    return ch-'0';
}

void setup()
{
    pinMode(ledPin, OUTPUT); ❽
    Serial.begin(9600); ❾
    Serial.println("Relay controller ready - www.BotBook.com");
}

void loop()
{
    if ( Serial.available()) { ❿
        char ch = Serial.read(); ⓫
        int pin = charToInt(ch); ⓬
        relayPulse(pin); ⓭
    }
    delay(20); ⓮
}
```

Let's take a look at the code:

❶   Use pin 13, which corresponds to the built-in LED, to indicate when the program holds the relay on.

❷   The relayPulse() function receives one integer as its parameter, which indicates which pin to pulse. For example, to pulse pin 2, you'd invoke the function as relayPulse(2).

❸   Because there are six controllable pins, set the output state so that you don't have to do this in the setup. This line is inside a function that we call over and over, so we'll be setting the output state more than once for each pin, but this has no harmful side effects.

❹   Now, set both the LED and the requested pin to HIGH (+5V).

❺   Wait for one second, which is a thousand milliseconds. The pin remains on for this duration of time.

❻ Switch off both the relay pin and the LED.

❼ Return a number corresponding to the character ch. The program converts from a character (char) to an integer using this formula ch-'0'. To understand this, suppose the character ch is '1'. In that case, the calculation would be '1'-'0'. In other words, you subtract the character 0's ASCII value from the ASCII value of the character 1. The ASCII value of character 0 is 48. The ASCII value of the character 1 is 49. Therefore, charToInt('1') returns the integer value 1 (49 - 48).

❽ Set the LED pin to OUTPUT mode so that it can be switched on and off.

❾ Open the serial port with the standard Arduino Serial Monitor speed of 9,600 bps.

❿ In the loop, check whether the serial port contains letters queued to be read. If so, execute the code in this block.

⓫ Read one character from the serial port and store it to the variable ch. The value of variable ch is of a character type, even though it will represent a number. For example, the character '2' is not the number 2; it is stored in the variable ch as an ASCII value (see *http://en.wikipedia.org/wiki/ASCII*). The ASCII value of character '2' is 50.

⓬ A character representing a number must first be converted into an integer. The function charToInt() takes care of this.

⓭ Perform a button press/release on the specified pin by calling the relayPulse() function.

⓮ Wait for 20ms, which equals 0.02 seconds—a very short amount of time. The delay keeps the Arduino's CPU from being taxed at 100% utilization (which would waste energy and subject the chip to excessive heat).

WHAT ARE THESE BLACK BOXES?

You might have noticed some components that we have not yet covered in the circuit board of the remote control. Many circuit boards have, among other components, black boxes with many pins. Those are *microchips*, a type of integrated circuit (IC; see Figure 7-22). A large number of components—such as resistors, transistors, diodes, and capacitors—have been integrated into one microchip. As its name implies, one microchip contains a circuit miniaturized into a very small size, designed for a specific purpose.

There are countless microchips for different purposes. Each chip has an ID code printed on top, which can be used to search for documentation on the manufacturer's website. Read the documentation and act accordingly.

Figure 7-22. Integrated circuits

Creating a Graphical User Interface

Typing numbers into the Arduino development environment's serial console to control power outlets from the computer is not a satisfying user experience. Clicking buttons in a graphical user interface is much more user-friendly.

As usual, before building the final program, we will test the features with smaller programs. It is best to make the test programs quite short and try only one feature at a time. We will therefore try the features of the program to be installed in the computer just like we have tested the sensors in previous chapters: first separately, and then as part of a device.

> *For information on installing and using Python on Windows, Mac, or Linux, see "Installing Python" in Chapter 5.*

Pack Many Buttons into One Window

You'll want many buttons in your remote control, so let's create a window with many widgets. The window is a container that can contain a button. Only one widget will fit in the window. If you try to add another widget to the same window, you'll get an error message, such as "A GtkWindow can only contain one widget at a time; it already contains a widget of type GtkButton." So we'll add one widget, but it will be a widget that can contain other widgets: a box that can contain multiple objects, buttons, or other widgets. The most important boxes are gtk.VBox and gtk.HBox.

The stacked widget box gtk.VBox gets the *V* in its name from the word *vertical*. In this type of box, items are packed vertically (stacked on top of each other).

The following program creates a new window, which has stacked buttons "Button 1" and "Button 2." Figure 7-23 shows the program in action.

```
#!/usr/bin/env python
# packingVBox.py - Pack many widgets on top of each other.
# (c) Kimmo Karvinen & Tero Karvinen http://BotBook.com

import gtk

window=gtk.Window()
window.connect("destroy", gtk.main_quit)

vbox=gtk.VBox()          ❶
window.add(vbox)         ❷

b1=gtk.Button("Button 1")    ❸
vbox.pack_start(b1)          ❹

b2=gtk.Button("Button 2")    ❺
vbox.pack_start(b2)

window.show_all()
gtk.main()
```

Let's review the function of the program line by line.

❶ Create a new gtk.Vbox class object.

❷ Add a box to the vbox window. After this, new widgets cannot be added to the window using the method add(). Instead, you'll add new widgets to the vbox.

❸ Add the b1 button into the box. The box was empty, so now the b1 button fills it.

❹ Pack the new button in the bottom of the box. The `pack_start()` method fills the vbox box from the bottom. This way, the button that was packed first will stay on top and the next one will appear below it.

❺ Create a new button, store it in the new b2 variable, and pack it in.

Figure 7-23. *Two buttons in one window*

Button Orientation

Using Hbox, you can place box widgets side by side. By packing widgets side by side with Hbox boxes and on top of each other with Vbox boxes, you can create any kind of layout.

The following program creates the combined Hbox and Vbox layout shown in Figure 7-24.

```python
#!/usr/bin/env python
# boxInsideBox.py - Combine boxes to lay out elements
# in rows and columns.
# (c) Kimmo Karvinen & Tero Karvinen http://BotBook.com

import gtk

window=gtk.Window()
window.connect("destroy", gtk.main_quit)

vbox=gtk.VBox()
window.add(vbox)

b1=gtk.Button("Button 1")
vbox.pack_start(b1) ❶

hbox=gtk.HBox()
vbox.pack_start(hbox) ❷

b2=gtk.Button("Button 2")
hbox.pack_start(b2) ❸

b3=gtk.Button("Button 3")
hbox.pack_start(b3) ❹

window.show_all()
gtk.main()
```

Here's how the code places a box within a box:

❶ Put the first button into the stacked widget's Vbox box.

❷ Add another box, an Hbox that places items side by side.

❸ Pack a button into the Hbox box.

❹ Add another button.

Now, you have a window with a "Button 1" button on top, and "Button 2" and "Button 3" buttons placed side by side beneath it.

Figure 7-24. *Buttons side by side and on top of each other*

Stretching Like Bubble Gum

Expand the window by clicking the lower-right corner and dragging it, as shown in Figure 7-25. Here, you can see one of the GTK library's strengths. When widget positions have been defined logically, and not as coordinates of pixels, windows are flexible. Using different settings, you can adjust whether the buttons are attached to walls or whether empty space appears in specific areas.

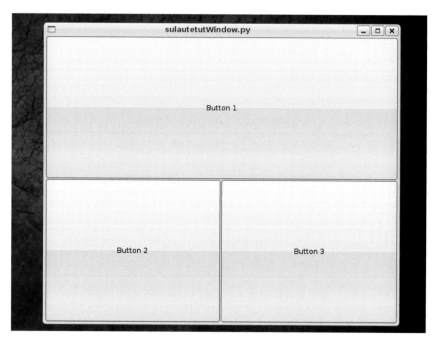

Figure 7-25. *Stretched window*

The Finished Remote Control Interface

The following program contains the completed remote control interface.

```
#!/usr/bin/env python
# remoteControl.py - Graphical user interface for remote control.
# (c) Kimmo Karvinen & Tero Karvinen http://BotBook.com

import serial ❶
import gtk

ser=None     # global variable ❷

def sendSerial(widget, ch): ❸
    global ser ❹
    print("Sending "+ch) ❺
    ser.write(ch) ❻

def main(): ❼
    global ser ❽
    # File name might differ between Linux, Windows, and Mac OS X ❾
    ser = serial.Serial('/dev/ttyUSB0', 9600)❿
```

```
    if (ser): ⓫
        print("Serial port " + ser.portstr + " opened.")

    window = gtk.Window(gtk.WINDOW_TOPLEVEL) ⓬
    window.connect("destroy", gtk.main_quit) ⓭

    vbox=gtk.VBox() ⓮
    window.add(vbox)
    row1=gtk.HBox() ⓯
    off1 = gtk.Button("1 off") ⓰
    off1.connect("clicked", sendSerial, "2") ⓱
    row1.pack_start(off1) ⓲
    vbox.pack_start(row1)
    on1 = gtk.Button("1 on") ⓳
    on1.connect("clicked", sendSerial, "8")
    row1.pack_start(on1)

    row2=gtk.HBox() ⓴
    vbox.pack_start(row2) # vertical box starts at the bottom
    off2 = gtk.Button("2 off")
    off2.connect("clicked", sendSerial, "3")
    row2.pack_start(off2)
    on2 = gtk.Button("2 on")
    on2.connect("clicked", sendSerial, "7")
    row2.pack_start(on2)

    row3=gtk.HBox()
    vbox.pack_start(row3)
    off3 = gtk.Button("3 off")
    off3.connect("clicked", sendSerial, "4")
    row3.pack_start(off3)
    on3 = gtk.Button("3 on")
    on3.connect("clicked", sendSerial, "6")
    row3.pack_start(on3)

    window.show_all() ㉑
    gtk.main() ㉒
    print ("Thanks for using BotBook.com remote control.")

if __name__ == "__main__": ㉓
    main()
```

Run the program and press the buttons shown in Figure 7-26.

Here's what's going on in the program, line by line:

Figure 7-26. *Button interface*

❶ Import two program libraries (serial and gtk). Read the serial port using the serial library and draw the graphical user interface window and buttons with the GTK library.

❷ Define the ser variable in the beginning of the program, outside of all functions. This way, it will become a global variable, which means it can be used by all functions. Initialize the ser variable with None, an empty value. In Python, variables are typed dynamically, so the type does not have to be described during the declaration of a variable. In this way, Python differs from, for example, the C language that Arduino is based on.

❸ The sendSerial method accepts two parameters, widget and ch. The first parameter is a widget object indicating which button was clicked (e.g., off1, on3, or off2). The other parameter, ch, is the character to be sent to the Arduino. For example, if the function is invoked as sendSerial(on1, .8.), ch takes on a value of character '8'.

❹ Use a global ser variable, visible in all functions. Normally, variables are visible only in the same block of functions in which they are declared. Global variables are an exception. Here, the global ser variable is an object you can use to write to the serial port opened earlier.

❺ Display a message to the console to help debug. For example, if you were sending the character '8', this would display "Sending 8".

❻ Write the ch character, such as '8', to the serial port.

❼ This is the main function of the program; the last line of this program will run main() when the program starts.

❽ Use the global ser variable, declared in the beginning of the program.

❾ This comment reminds you that you need to change the name of the serial port (*/dev/ttyUSB0*) to the name of the serial port Arduino is connected to, such as *COM1* on Windows or something like */dev/tty.usbserial A700dECp* on the Mac. In the Arduino IDE, you can confirm which serial port you are using by clicking Tools→Serial Port. For more information on Arduino and serial ports, see "Hello World with Arduino" in Chapter 2.

❿ Open a connection to the serial port, using 9,600bps as the speed. As the initial capital letter indicates, serial.Serial is a class. Here, the program calls the Serial class constructor, which returns a Serial class object called ser.

⓫ If the ser object was created successfully, display this text to confirm it.

If the program can't open the serial port, Python will crash the program and automatically display an error message before it terminates. The most common error is trying to open the wrong serial port. Here's an example of such an error:

```
serial.serialutil.SerialException: could not open port /dev/ttyUSB1: [Errno
2] No such file or directory: '/dev/ttyUSB1'
```

You can correct the error by specifying the correct serial port, as explained in Step 9.

⓬ Create a new window object called window. This is the main window of the program.

⓭ This gives the user a meaningful way of closing the program, connecting the window-closing button (usually an X) to the gtk.main_quit() function. This function ends the main loop. Later on in this function, you'll see a call to gtk.main(). After that, the program will display only "Thanks for using BotBook.com remote control", so ending the loop will end the whole program.

⓮ A window can contain only one widget, such as a button. The interface needs six buttons, so the program must use boxes (box) to create a stacked widget box (vbox) to add to the window.

⑮ Create a stacked Hbox widget box called row1 and put it into the stacked VBox widget box. Now, the interface has only one row in the stacked widget's vbox. The row has been divided into parallel positions with an Hbox named row1.

⑯ Create a new button with the text "1 off" and name it off1.

⑰ Connect the button's click event to the sendSerial() function. Clicking the off1 button sends it the clicked event, after which the button calls sendSerial(off1, .2.). (According to Table 7-1, pin 2 corresponds to the Off 1 button.)

⑱ Pack the button into the row1 box, and the row1 box into the vbox.

⑲ Create the remaining buttons in the same way and pack them in as the program continues.

⑳ Create a new parallel widget's HBox called row2 and pack it in the stacked widget's box. Keep adding buttons and one more HBox (row3) for the last row before finishing.

㉑ Present the user interface to the user. In addition to showing the window object, the method show_all() shows all the elements the window contains. This makes all the boxes and buttons visible.

㉒ Invoke gtk's main loop, gtk.main(). Most of the time is spent in the main loop, waiting for the user to do something.

㉓ Invoke the main() function when the program runs.

When a button, such as "1 on," is clicked on the computer, the Python program sends the corresponding pin number (8, as shown in Table 7-1) to the serial port.

The sketch running on the Arduino switches that pin (8) on for one second. A relay connected to that pin switches on the corresponding remote control button for one second. The remote control sends a message wirelessly to the receiver (AC socket), after which it is switched on.

Refer back to Figure 7-26 to see the remote control's user interface running on a computer.

Creating an Enclosure

Next, you will create an enclosure for the remote control so it can be stored next to the computer. Once you've finished, you can go on to develop more complex and intelligent ways to use the controller.

This project creates an enclosure from a freezer storage container (Figure 7-27), which works well for packing prototypes of different sizes and shapes. These boxes are also inexpensive and easy to cut and drill.

Create some small holes in the box in order to attach the remote control's circuit board. The circuit board will already have suitable holes where it was attached to its original housing (and you should have the screws left over from it as well). Screw the circuit board into place and attach a round sticker on the

Figure 7-27. *Our enclosure is created from a freezer box*

outside of the box over the green light (most remote controls have an indicator light that shows when you pressed a button). The sticker ensures that the light will still show through the case after you've painted it. Figure 7-28 shows the screws, controller board, and sticker in place.

Remove the circuit board and block the holes from inside with tape, as shown in Figure 7-29. Close the lid, and you are ready to paint the case.

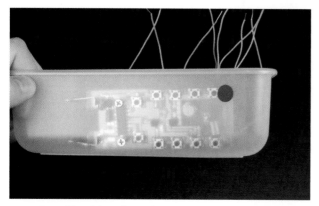

Figure 7-28. *Circuit board attached, with a round sticker over the LED indicator light*

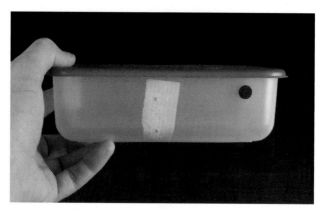

Figure 7-29. *Enclosure ready to be painted*

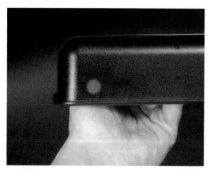

Figure 7-30. *Round sticker removed*

Spray paint the box in several thin layers. When the painting is done, remove the round sticker and paint one more thin layer over the area. Now the surface looks dark when the light is not on, as shown in Figure 7-30, but the light will still show through when the remote control sends data.

Screw the circuit board back in place. Press the prototyping breadboard with its components in the bottom of the box. The board fit perfectly in the box we used, the pressure of the box's walls holding it solidly in place. If you're not as lucky, you can hot-glue or screw the board into place.

You can also attach felt pads to the bottom of the box, as shown in Figure 7-31, so that it won't scratch a table or a computer you might place it on later. Figure 7-32 shows the remote control and the breadboard circuit in place.

Figure 7-31. *Felt pads attached to the bottom*

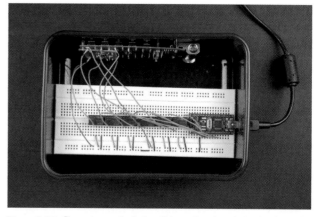

Figure 7-32. *Remote control circuit board and prototyping breadboard with its components in place*

When the board is in place, use a pen to trace the Arduino USB port and then drill a hole to pull the cable through (see Figures 7-33 and 7-34). Move the Arduino away from the area of the hole to avoid breaking the chip.

Figure 7-33. *Drilling a hole for the USB cable*

Figure 7-34. *Finished hole*

You're finished! Figure 7-35 shows the cable in place, and Figure 7-36 shows the finished project.

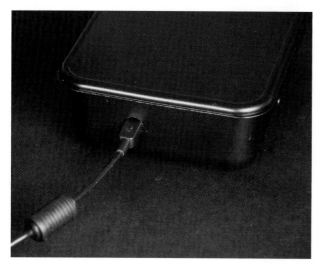

Figure 7-35. *Cable in place*

Figure 7-36. *Finished remote for a smart home*

Soccer Robot

8

The audience is cheering; the opposing team has fallen behind. You control your player—a Soccer Robot—by tilting your cell phone in the air like a steering wheel. Stop, kick, and gooooooooooal! This project combines the techniques you've learned in previous chapters to make this game a reality.

In this chapter, you will learn how to make Arduino communicate with an Android cell phone wirelessly via Bluetooth. You'll measure the position of the cell phone with its accelerometer. First, though, you'll build a sturdy frame for your Soccer Robot, a moving robot on wheels (shown in Figure 8-1, Figure 8-3, and Figure 8-4). The robot's structure adapts well to other projects, because you can build it easily and add many sensors and accessories to it.

> *This chapter includes examples for Android devices. The sample code, as well as sample code for Nokia Symbian devices, is available from* http://BotBook.com/. *Check this book's catalog page at* http://oreilly.com/catalog/0636920010371 *for an upcoming ebook with a detailed look at the Nokia Symbian example code.*

You can invent all types of functions for the moving robot on wheels. After completing this project, you could turn your Soccer Robot into an autonomous version capable of avoiding obstacles, or teach it to follow lines with QTI line sensors. You can also combine the cell phone control and autonomous functionality into, for example, a robot that wanders around a room until you alter its behavior with the press of a button.

> *Before you attempt this project, be sure you've gone through one of the earlier Arduino projects and the mobile phone Boxing Clock project in Chapter 6. You'll be combining your Arduino and mobile phone programming experience in this chapter. If you take things one step at a time, you'll find working with this project much easier and more fun.*

Figure 8-1. *The Soccer Robot*

199

What You Will Learn

In this chapter, you'll learn how to:

- Build a frame for a wheeled robot

- Disassemble parts from a hard disk and other things for reuse

- Control Arduino from an Android cell phone via Bluetooth

Tools and Parts

You'll need the following tools and parts (shown in Figure 8-2) for this project.

Manufacturer part numbers are shown for:

- Maker SHED (US: http://maker-shed.com): SHED

- Element14 (International and US; formerly Farnell and Newark, http://element-14.com): EL14

- SparkFun (US: http://sparkfun.com): SFE

1. Two small, metal strips for the front fork. You could salvage these from other devices, such as an old typewriter. If you have metal snips and a small amount of sheet metal, you could also cut them yourself (but be sure to use a metal file or metal sandpaper to smooth the edges, which will be extremely sharp).

2. Metal strips for the servo attachments (also salvaged or cut to size from sheet metal).

3. Three servo extension cables (SFE: ROB-08738).

4. Two small springs.

5. An LED (EL14: 40K0064; SFE: COM-09592).

6. Twenty-four 3mm screws and nuts.

7. Jumper wires in three colors (SHED: MKEL1; EL14: 10R0134; SFE: PRT-00124).

8. AC lighting connector block (RadioShack sells it as a "*European-style terminal strip*").

9. Broken hard drive (this will be dismantled).

10. Hot-glue gun and hot-glue sticks.

11. Electric drill.

12. Battery box for two AA batteries (for Arduino BT; SFE: PRT-09547) or four AA batteries (Arduino Pro Mini with Bluetooth Mate; SFE: PRT-00552).

13. Three servo motors: one normal servo (SFE: ROB-09064) and two continuous rotation servos (SFE: ROB-09347).

14. A piece of metal suitable for kicking the ball.

15. Wire strippers (EL14: 61M0803; SFE: TOL-08696).

16. Diagonal cutter pliers (EL14: 52F9064; SFE: TOL-08794).

17. Hammer.

18. Metal saw.

19. Two wheels.

20. Furniture wheel with ball bearings.

21. Arduino with Bluetooth module.

22. Pliers.

23. Heat-shrink tubing.

24. Two felt pads.

25. Two metal drill bits (3mm and 6mm).

26. Tape measure.

27. Soldering iron and solder.

28. USB cable compatible with the cell phone.

29. Cell phone with Android 2.1 API level 7, Bluetooth, and an accelerometer. We have built this project with the Google Nexus One and the Sprint EVO 4G. (We've even built it with a Nokia phone, but these instructions are only for Android.)

30. Another set of pliers.

31. Marker.

32. Center punch.

33. Metal spike.

34. Screwdriver with Phillips (cross-head), flat-head, and Torx bits that can open the screws on the hard drive.

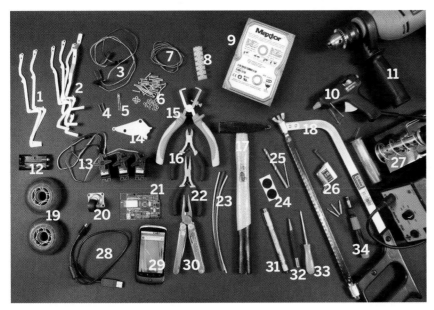

Figure 8-2. *Parts needed for this chapter*

WHICH ARDUINO?

The original version of this project used the Arduino BT shown in Figure 8-2. However, the BT is expensive and fragile, so we recommend using a newer Arduino model.

One advantage of the BT is that it can run on only 3 volts. So, we suggest you use a 3.3V Arduino: specifically, SparkFun's Arduino 3.3V Pro Mini (DEV-09220; *http://www.sparkfun.com/products/9220*) and connect it to the SparkFun Bluetooth Mate Silver (WRL-10393; *http://www.sparkfun.com/products/10393*).

At $60, this is also a less expensive option ($20 for the Arduino Pro Mini + $40 for the Bluetooth Mate, rather than $150 for the Arduino BT).

Like some other low-profile Arduino boards (and clones), the Pro Mini does not include a USB-TTL serial converter. This means you can't upload an Arduino sketch to it without a USB-TTL breakout board such as the SparkFun FTDI Basic Breakout – 3.3V (DEV-09873; *http://www.sparkfun.com/products/9873*). One reason the Pro Mini ends up being cheaper than full-size Arduinos in the long run is that you can buy a handful of Pro Minis and program them all with the same FTDI breakout board.

> *Another advantage of the Pro Mini is that the Bluetooth Mate plugs into the same pins that the USB-TTL serial converter uses. This means that when you are debugging, you can plug the USB-TTL serial converter into the Pro Mini instead of the Bluetooth Mate, and use the Arduino Serial Monitor or a serial terminal program (see "Testing the Bluetooth Connection," later in this chapter) to type commands to the Pro Mini, mimicking commands that were sent over a Bluetooth connection.*

Figure 8-3. *Kimmo Karvinen's and Mikko Toivonen's original Basic Stamp version of the Soccer Robot at the University of Art and Design Helsinki's Demo Day in 2008*

Figure 8-4. *Another view of the original robot*

Continuous Rotation Servos

In Chapter 4, you learned how to control servo motors, whose motion is constrained to rotary motion at specific angles. Because this project will use two servos for rotating the wheels, the servos must be able to rotate freely; the Parallax continuous rotation servos pictured in Figure 8-5 are a good example.

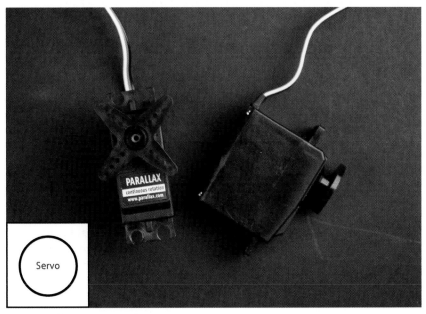

Figure 8-5. *Continuous rotation servos*

Controlling Continuous Rotation Servos

Connect the servo motor's black wire to the Arduino's GND pin and the red wire to the +5V pin. Connect the white wire to digital pin 2 to control the motor. Figure 8-6 shows the connection diagram.

Not all servo motors are calibrated identically, so you might need to tweak the values you send from Arduino to move the rotor. Some continuous rotation servos have an adjustment screw, which changes its center point (the point at which the rotation stops). If your motor does not have such a screw, you can use the code below to find the right center point (which is normally a value of 90 degrees) and modify the example programs in this chapter accordingly. Use a value that keeps the servo steady. Later in the project, when the servo spins in different directions, use values that move your servo to the left or right at the desired speed. Chapter 4 has more information on working with servos.

Before you adjust the screw, upload the servo-centering code in the following section to the Arduino.

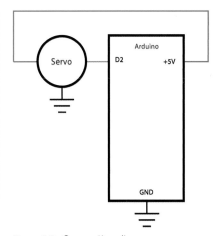

Figure 8-6. *Connection diagram*

Centering (Stopping) the Servo

If your servo has an adjustment screw, it's easy to center it. Just run this code to keep the servo turned to 90 degrees. Then adjust the screw until the servo stops moving.

```
// centerFullRotation.pde - Halt a continuous rotation servo
// (c) Kimmo Karvinen & Tero Karvinen http://BotBook.com

#include <Servo.h>

Servo myServo;
int servoPin = 2;

void setup()
{
  myServo.attach(servoPin);
}

void loop()
{
  myServo.write(90);
  delay(20);
}
```

The program writes 90 (degrees) to the servo. This centers the servo and holds it steady. We call `write()` repeatedly inside of `loop()`, even though the servo will hold steady after calling `write(90)` only once. In the next sketch, you'll see how to vary the speed within `loop()`.

Finding the Center Point

Few continuous rotation servos have a centering or adjustment screw. What if your servo doesn't have one? You can modify your Arduino code to send the suitable angle to stop the servo.

For example, it could be that `myServo.write(90)` makes the servo turn slowly, and you have to send `myServo.write(94)` to make it stop. With the following code, you can determine the correct stopping angle; it slowly increases the servo's turn from 0 degrees (full speed ahead) to 180 degrees (full speed backward) and writes the current value to the serial port.

Once the servo stops, you have found the center point. Read the value from the Serial Monitor in the Arduino IDE and write it down.

```
#include <Servo.h>

Servo myServo;
int servoPin = 2;

void setup()
{
  myServo.attach(servoPin);
  Serial.begin(115200);   // bit/s ❶
}
```

```
void loop()
{
  for (int i=0; i<=180; i=i+1) {  ❷
    myServo.write(i);  ❸
    Serial.println(i);
    delay(150);  ❹
  }
}
```

❶ Initialize the serial connection between the Arduino and the computer. You can read the messages from the Serial Monitor in the Arduino IDE. Remember to set the same speed in both the Arduino code and the Serial Monitor on the computer.

❷ Initially, loop variable i is declared and is set to 0.

The condition is checked, and because 0<=180, we enter the first iteration. We run the contents of the block: the code inside the curly braces ({and }). At the end of the first iteration, we increment the loop variable i by one, so i is now 1. This completes the first iteration of the loop.

The second iteration starts with checking the condition, and 1<=180. Then we run the contents of the block and increment the counter by one. This goes on for many iterations.

Finally, i equals 180, and we run the loop for the last time. When the counter is incremented to 181, the loop condition 181<=180 is false, and the loop ends.

❸ The servo is turned to the position shown by the loop variable i. In the first round, it's turned to 0 degrees. In the last round, it's turned to 180 degrees.

❹ Wait for 150ms, or 0.15s, so that you have time to read the value before the servo moves again. If you need more time to read the value, you can increase this number.

Rotating to Different Directions

With a standard servo, the value you pass to Servo.write() specifies the angle (in degrees) in which the servo should move. With a continuous rotation servo, this value instead specifies the *speed*:

0
> Full speed in one direction

89
> Slowest speed in that direction

90
> Stopped

91
> Slowest speed in the other direction

180
> Fastest speed in that direction

This sketch rotates the servo clockwise, holds it steady for a second, rotates it counterclockwise, stops, and then does it all over again:

```
// fullRotation.pde - Turn continuous rotation servo clockwise,
// counterclockwise and stop.
// (c) Kimmo Karvinen & Tero Karvinen http://BotBook.com

#include <Servo.h> ❶

Servo myServo; ❷
int servoPin = 2; ❸

void servoClockWise() ❹
{
    myServo.write(85); ❺
    delay(1000); ❻
}

void servoCounterClockWise()
{
    myServo.write(95); ❼
    delay(1000);
}


void servoStop()
{
    myServo.write(90); ❽
    delay(1000);
}

void setup()
{
  myServo.attach(servoPin); ❾
}

void loop() ❿
{
  servoClockWise();
  servoStop();
  servoCounterClockWise();
  servoStop();
}
```

Let's look at the code line by line:

❶ Pull in the Servo library.

❷ Define a Servo object named myServo.

❸ Use pin 2 to control the servo.

❹ We created functions for each action we want to perform. In the loop() function, we stack these functions one after the other just like building blocks. All the functions are quite similar. Like the others, this one takes no parameters.

❺ Rotate the pin in a clockwise motion, using the Servo library command. Because it's in a loop, it keeps rotating for as long as it takes the loop to complete.

❻ Keep going for one second.

❼ Rotate in the other direction.

❽ Stop the servo.

❾ Attach the `Servo` object to the specified pin.

❿ We'll keep rotating the servo back and forth until you pull the plug on the Arduino.

You can admire the rotation of your servo for a while. Can you program other movements for it?

Modding a Standard Servo into a Continuous Rotation Servo

Continuous rotation servos are harder to find than standard ones, and they're often more expensive. But you can modify a standard, limited rotation servo to make it a continuous rotation model.

Certain servo brands are easier to modify than others. Some manufacturers, such as Parallax, design their servos for ease of modification. You can find their instructions at *http://www.parallax.com/dl/docs/books/edu/roboticsservomod .pdf*.

To modify any other model of standard servo, do the following:

1. Remove the horn from the servo (it is usually held in place by a screw).

2. Remove the four screws from the bottom of the servo, as shown in Figures 8-7 and 8-8, and set them aside.

> *One limitation to this hack is that servos modified this way do not have an adjustment screw, which means that you might have to tweak your code to calibrate the servo. For example, you might need to change the 90 in* `myServo.write(90)` *to a value you arrive at by trial and error. If you find that* `myServo.write(93)` *or* `myServo.write(88)` *stops the servo, use that instead.*

Figure 8-7. *Screws to be removed*

Figure 8-8. *Loosening the screws*

3. Remove the cover of the servo with the bottom down to prevent the gears from falling out. Make a note of how the gears are arranged (take a couple of photos with a digital camera if you need to).

4. Next, you must remove the obstructions that prevent continuous rotation. Most models have obstructions in both the gearing and the top lid of the servo. The gears in Figure 8-9 also have a small screw that you must remove before cutting off the obstructions.

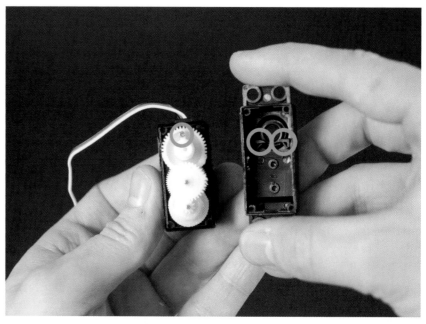

Figure 8-9. *Obstructions that block rotation*

5. Cut or snap off pieces that are in the way, as shown in Figure 8-10. You can make the first cut with an X-ACTO knife or angled cutters. You might want to clean up the areas with a small file or a mini drill. Put the lid in place and try to rotate the motor from the servo horn. If it does not turn smoothly all the way around, open the servo and even out the areas better. Figure 8-11 shows the servo with the obstructions removed.

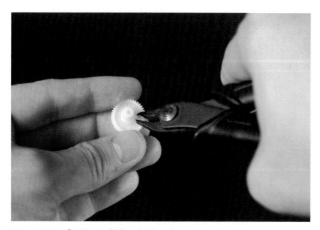

Figure 8-10. *Cutting off the obstructions*

Figure 8-11. *Obstructions removed*

6. Next, you must prevent the servo from being able to determine its position. Lift off the gears. Under one of the gears, you'll find the shaft of a potentiometer (see Figure 8-12), which the servo uses to measure its position.

Figure 8-12. *Shaft of the potentiometer*

7. The gear you removed from above the potentiometer has an indentation into which the potentiometer's shaft locks, as shown in Figure 8-13. Even out the edges of this dent to make it perfectly round, as shown in Figure 8-14, after which the potentiometer's shaft will no longer rotate when the gear turns. (Another option is to cut off the potentiometer's head, but that is harder.)

Figure 8-13. *Area of the gear onto which the potentiometer's shaft locks*

Figure 8-14. *Rounding out the dent in the gear*

8. Using small needlenose pliers, rotate the shaft of the potentiometer right and left to determine its range of movement. Turn it to its center position, as shown in Figure 8-15. This way, the servo thinks that it is always centered, and it will be able to rotate continuously. To make sure it's calibrated,

connect the servo to Arduino and run the centering sketch shown earlier in "Centering (Stopping) the Servo." When the servo stops moving, you've calibrated it correctly.

Figure 8-15. *Adjusting the potentiometer to the center*

9. To secure the potentiometer, add a drop of hot glue on top of it (see Figure 8-16). Make sure you use only a small amount of glue, so the other gears can still move freely.

Figure 8-16. *Using a drop of hot glue to hold the shaft of the potentiometer in place*

10. Put the gears back in place and screw the lid back on. Now you have a continuous rotation servo, as shown in Figure 8-17.

Figure 8-17. *Finished continuous rotation servo*

Connecting the Arduino to the Bluetooth Mate

If you're using a SparkFun Arduino Pro Mini, you can plug the Bluetooth Mate directly into the Arduino. The Bluetooth Mate does not come with any headers soldered on, so you'll need to solder on a set of female headers to plug them into the Arduino. Figure 8-18 shows the Bluetooth Mate connected to the Arduino Pro Mini with a battery pack.

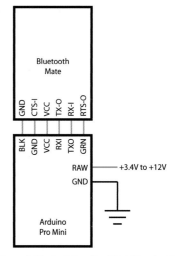

Figure 8-18. *Bluetooth Mate connected to Arduino Pro Mini*

If you're using another model of Arduino, you can solder a set of male headers to the Bluetooth Mate, plug it into a breadboard, and use jumper wire to connect it to the Arduino by following the connection diagram shown in Figure 8-19.

To power the project on its own, you need to connect a 3.3V power supply to the Arduino. Attach the black wire of the battery compartment to the Arduino GND connector. Connect the red wire of the battery compartment to the Arduino VCC (Arduino Pro Mini) or Vin (Arduino BT) connector. Place two AA batteries in the battery compartment only after you've completed the connection. If you're using the Arduino Pro Mini, you will find it helpful to build this project on a half-size breadboard, as shown in Figure 8-19.

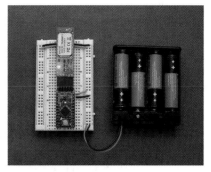

Figure 8-19. *Arduino Pro Mini, Bluetooth Mate, and power supply connection diagram*

> *Two AA batteries aren't capable of delivering 3.3 volts to the Arduino. Two fresh batteries will deliver 3 volts, and this will gradually drop as the charge runs down. Some rechargeables deliver only 1.2 volts each, so you're just giving the project 2.4 volts at first. In our testing, this project ran for a couple hours on 2 AAA alkaline batteries. However, for long-term usage, it would be better for you to use four AA rechargeables (4.8–6 volts, and with greater capacity than AAA) and connect the red wire to the Arduino Pro Mini's RAW pin instead of VCC. This sends the current through the Pro Mini's onboard voltage regulator, dropping it to 3.3 volts and making everyone (the Arduino and the Bluetooth module) happy.*
>
> *The Arduino BT is much more content with a varying range of low voltages (1.2–5.5 volts) than the Pro Mini, which requires 3.35–12 volts on the RAW (regulated) pin, or 3.3 volts on the VCC (unregulated) pin. Don't send more than 3.3 volts into the VCC pin, or you might burn out the Pro Mini.*

Switch on the power. Now you're ready to connect to the Bluetooth module from a computer. To do this, you'll need a Windows, Mac, or Linux system with a Bluetooth module. Many computers come with Bluetooth built in, but you can easily find low-profile USB Bluetooth modules such as IOGear's Bluetooth Micro Adapter or Belkin's Mini Bluetooth Laptop Adapter.

> *Just before you try connecting from a computer, turn off power to the Arduino and the Bluetooth Mate, and then turn it back on. This will ensure that the Bluetooth Mate is discoverable.*

Windows 7

Find the Bluetooth icon in the System Tray (also known as the Notification Area). You might need to click Show Hidden Icons in the System Tray before you see the Bluetooth icon. If you do not see it, then your computer is not configured properly for Bluetooth or does not have Bluetooth built in. Next:

1. Click the Bluetooth icon and select "Add a device."

2. Choose the FireFly or RN42 device (it might initially appear as Other), as shown Figure 8-20, and click Next. FireFly or RN42 is the name that the Bluetooth Mate uses to identify itself (because it uses a Bluetooth chip from Roving Networks' FireFly line of products).

3. When asked to select a pairing option (see Figure 8-21), click "Enter the device's pairing code" and type **1234** when prompted.

4. In the final window that appears ("This device has been successfully added to the computer"), click "Devices and Printers."

5. Locate the FireFly/RN42 device in the list, and double-click it to bring up its properties. On the Hardware tab, you can see which COM port it is using, as shown in Figure 8-22. Make sure you remember this COM port, because you'll need to use it later in "Testing the Bluetooth Connection."

Figure 8-20. *Selecting the Bluetooth Mate (FireFly) from the list of Bluetooth devices*

Figure 8-21. *Providing a passkey (the default is 1234)*

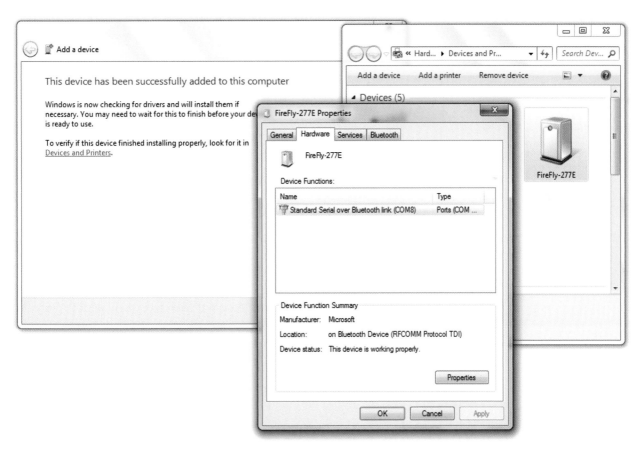

Figure 8-22. *Bluetooth Mate settings (make a note of these somewhere)*

Ubuntu Linux

Locate the Bluetooth icon on the right side of the GNOME desktop panel at the top of the screen. Click it and choose "Set up new device." Then:

1. Click Forward. Choose the FireFly or RN42 device (it might initially appear as Unknown).

2. Click Pin Options, choose 1234, and click Close.

3. Click Forward.

4. Click Close when you're finished.

Now you need to set up a serial port to use with it. This will be much easier if you install the blueman Bluetooth Manager:

1. Choose System→Administration→Synaptic Package Manager (you'll need to supply your password).

2. Search for blueman, and when it appears in the list, click its checkbox and choose "Mark for Installation."

3. Click the Apply icon, review the dialog that appears, and click Apply to start the installation.

4. Exit the Package Manager when you're done.

Each time you want to connect to the device from your computer, fire up the Bluetooth Manager by clicking System→Preferences→Bluetooth Manager.

Locate the FireFly or RN42 device in the list, right-click it, and choose Serial Service or SPP under "Connect To:". Once it's connected, a message will appear, saying something like "Serial port connected to /dev/rfcomm0." Make a note of this serial port, because you'll need to use it later in "Testing the Bluetooth Connection."

> If you are unable to connect to the RFCOMM port later (specifically, if you get an error that the device is busy or that access is denied), try toggling the port off (right-click FireFly and choose "Serial Port rfcommX" under "Disconnect:") and then connect to it again.
>
> Do this a couple of times, and you might be able to trick Ubuntu into giving you access. For up-to-date information on the bug behind this, see https://bugs.launchpad.net/ubuntu/+source/linux/+bug/570692.

Mac OS X

Click the Bluetooth menu icon on the right side of the Mac OS X menu bar (it's usually next to the WiFi icon). From the menu that appears, choose "Set up Bluetooth Device." Then:

1. Wait for the Bluetooth Setup Assistant to detect the Bluetooth Mate (it's listed as an RN42 or FireFly device, as shown in Figure 8-23). Make a note of the name, because it will be used to name your serial port. For example, the device shown here is FireFly-277E. The serial port that's created by this procedure is in the format /dev/tty.DEVICENAME-SPP, so in this case the serial port is /dev/tty.FireFly-277E-SPP. You'll need to use this serial port next in "Testing the Bluetooth Connection."

2. Mac OS X will try to pair using the passkey 0000, which will fail (Figure 8-24). Click Passcode Options.

3. Select the "Use a specific passcode" option, type **1234** as shown in Figure 8-25, and then click OK. Now you can complete the setup process.

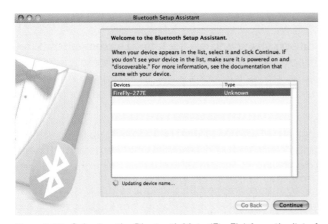

Figure 8-23. *Selecting the Bluetooth Mate (FireFly) from the list of Bluetooth devices*

Figure 8-24. *The default passkey failing*

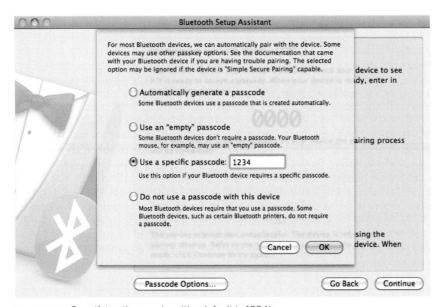

Figure 8-25. *Specifying the passkey (the default is 1234)*

Testing the Bluetooth Connection

Upload the following program to your Arduino. You will need to temporarily disconnect the Bluetooth Mate, disconnect the Arduino from battery power, and plug the Arduino into your computer using the FTDI breakout board (or a compatible FTDI cable).

This program waits for commands over the serial port and echoes the commands back so you can be sure that it received them. 1 turns on pin 13, and 0 turns it off. Like other Arduinos, the Arduino Pro Mini has an LED connected to pin 13, so you can watch it blink.

```
/*
 * BlinkOnCommand - turns an LED on or off
 */
```

```
int ledPin = 13;

void setup() {
  Serial.begin(115200);    // Open a connection to the Bluetooth Mate
  pinMode(ledPin, OUTPUT); // Activate the LED
}

void loop() {

  // Look for data coming in from the Bluetooth Mate
  if (Serial.available() > 0) {
    char cmd = Serial.read(); // Read the character
    Serial.print(cmd);        // Echo the character back

    // '1' turns on the LED, '0' turns it off
    if (cmd == '1') {
      digitalWrite(ledPin, HIGH);
    } else if (cmd == '0') {
      digitalWrite(ledPin, LOW);
    }
  }
  delay(20);
}
```

After you upload the program, reconnect the Bluetooth Mate and connect the battery pack, but don't turn it on yet.

Next, you'll need a serial terminal program such as CoolTerm (Mac and Windows: *http://freeware.the-meiers.org/*) or PuTTY (Windows or Linux: *http://www.chiark.greenend.org.uk/~sgtatham/putty/*). On Mac or Linux, you can also use the built-in program *screen*, which you can run from the command line.

A serial terminal program is a lot like Arduino's built-in Serial Monitor, except a serial terminal program lets you have an interactive session with another device: the device can respond to keystrokes as soon as you type them.

Although a serial terminal program is typically used to communicate with a device connected by a serial cable, it also works with wireless serial cable replacements such as Bluetooth. So, instead of specifying the name of a physical serial port (such as COM1 on Windows, */dev/ttyUSB0* on Linux, or */dev/tty.usbserial-FTD61SRE* on the Mac), you'll give the serial terminal program the name of the serial port you obtained back in "Connecting the Arduino to the Bluetooth Mate."

Here's how to connect on various terminal programs. Power on the Arduino and Bluetooth Mate, and then follow these instructions:

CoolTerm
> Click the Options button. In the dialog that appears, choose the serial port you set up in "Connecting the Arduino to the Bluetooth Mate." Set the Baudrate to 115,200, and click OK. Then click Connect.

PuTTY
> Under Connection Type, choose Serial. Then type the name of the serial port you set up in "Connecting the Arduino to the Bluetooth Mate." Set the speed to 115,200, and then click Open.

screen

> Open a Terminal Prompt (Linux: click Applications→Accessories→Terminal; Mac: open the *Applications* folder, go into the *Utilities* subdirectory, and double-click Terminal). Run the command **screen /dev/*PORT* 115200** (for example, `screen /dev/rfcomm0 115200` or `screen /dev/tty.FireFly 277E-SPP`). Press Ctrl-A, K to quit screen.

You should see the Bluetooth Mate's green light come on, indicating that you're connected. Type **1** to turn the LED on and **0** to turn it off.

Building a Frame for the Robot

A hard drive's casing is a fantastic material to use for building the robot frame. It's lightweight, stiff, practically free, and you can drill or saw into it. An added bonus is the street cred you get from repurposing broken or obsolete hardware.

Open up the Torx screws in the cover of the hard drive (Figure 8-26) and remove the top (Figure 8-27). Some of the screws are usually hiding underneath stickers. Remove the remaining screws from the inside so that you can use the rest of the cover as a base for future projects. It is also wise to remove and save the strong magnets, because you can use them for all sorts of purposes (such as picking up keys you've dropped down a drain).

> If you have any problems connecting, turn the Arduino and Bluetooth Mate off and on again and try these steps right away. For more troubleshooting help, check the comments on the SparkFun's product page (http://www.sparkfun.com/products/9358) or SparkFun's Wireless/RF discussion forum (http://forum.sparkfun.com/viewforum.php?f=13).

Figure 8-26. *Locations of screws*

Figure 8-27. *Hard drive opened*

Mark the places for the holes (Figure 8-28) with a marker. Make sure that the holes for the servos and rear wheel will fit to the parts you are using. Use a center punch to make indentations where the holes will go (Figure 8-29). (Don't skip this step; drilling with a hand drill is hard unless you've started the holes.)

Figure 8-29. *Making marks for the holes with a punch*

Figure 8-30. *Drilled 3mm holes and notch cut for the kicker leg*

Figure 8-31. *Bent metal strips*

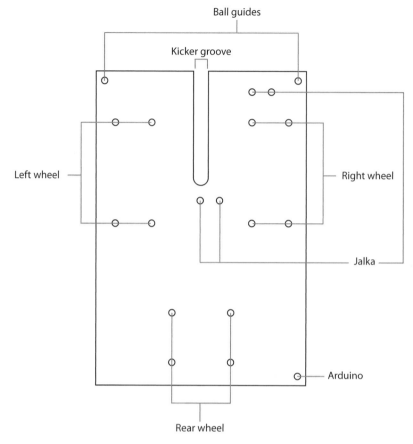

Figure 8-28. *Locations of holes*

Using a metal drill bit, drill 3mm holes where you started them with the punch. You will also need to make a notch for the leg. Drill a 6mm hole at the end of the groove and cut from there to the edge with a metal saw. Figure 8-30 shows the case with the holes drilled and a notch cut.

Making the Servo Attachments

To make the servo attachments, you'll need some metal strips; you can probably scavenge them from a broken piece of equipment such as a typewriter. If you don't have something suitable, purchase some L-brackets from a hardware store.

Cut six 5cm metal strips. Bend a right angle in each strip around the 3cm point, as shown in Figure 8-31. You've formed the attachments. (Although Figure 8-31 shows holes drilled into the attachments, don't drill them until you are attaching them to the servos in the next steps.)

If you bend the metal first in one direction and then back, it could break, so make the 90-degree bend in one movement.

You'll use two of the pieces as attachment parts for the kicker leg. Connect the longer part (the 3cm side of the bend) to the servo, as shown in Figures 8-32 and 8-33, separating it a bit from the frame. Drill holes in the attachments, lining them up with the holes in the frame and the servos, and use a bolt and nut to hold each attachment in place.

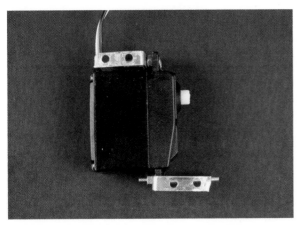

Figure 8-32. *Kicker servo attachments from the side*

Figure 8-33. *Kicker servo attachments from above*

Four pieces will form the attachments for the tires. Attach the shorter part (the 2cm long side of the bend) to the servo, as shown in Figures 8-34 and 8-35, and the longer part (the 3cm side of the bend) to the frame.

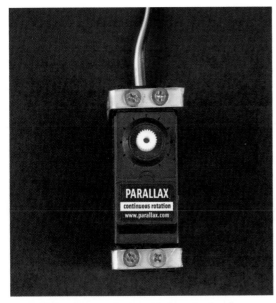

Figure 8-34. *Wheel servo attachments from the front*

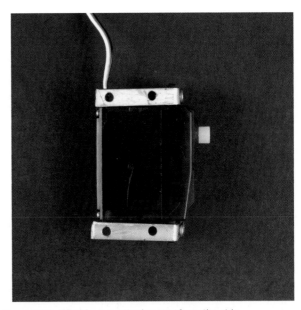

Figure 8-35. *Wheel servo attachments from the side*

Figure 8-36 shows all three servos with the attachments in place.

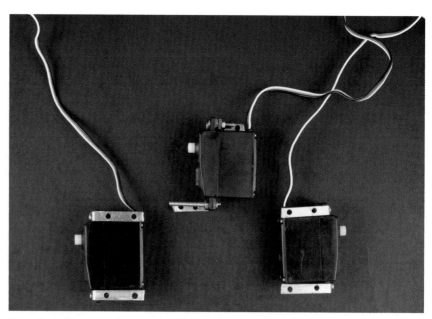

Figure 8-36. *Servo attachments placed as they attach to the frame of the robot*

Screw and bolt both wheel servos so that the side from which the wires come out points forward. Figures 8-37 and 8-38 show the wheel servos in place. You'll attach the kicker servo later.

Figure 8-37. *Attached wheel servos from above*

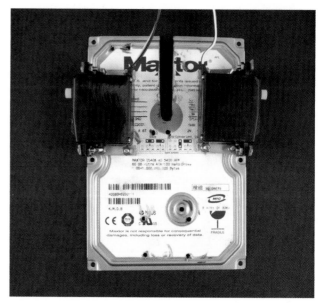

Figure 8-38. *Attached wheel servos from below*

Making the Wheels

You can find ready-made wheels to attach to servos (see *http://www.solarbot ics.com/motors_accessories/wheels/* for some inexpensive options), but you can also make them yourself. You can attach an object of almost any shape (but let's stick with a wheel here) by drilling a few holes in the object and screwing it to the servo horn. Figures 8-39 and 8-40 show a few possibilities: vacuum cleaner wheels, remote-control-car tires, and used rollerblade wheels. When choosing a suitable wheel, make sure that it is not overly heavy and that its surface has sufficient grip.

Figure 8-39. *Various wheels*

Figure 8-40. *Wheels from rollerblades*

We chose rollerblade wheels for our robot. We drilled the outermost holes of the servos large enough for the 3mm screws to fit through them. The roller-blade wheels already had holes in them, so we simply pushed screws through them to attach them to the horn of the servo.

Press the wheels to the servo and tighten the servo horn screw in place, as shown in Figure 8-41.

Attaching the Rear Wheel

You can use a furniture wheel (Figure 8-42) with ball bearings as an inexpensive and very functional rear wheel. It will follow the movements of the front wheels and allow rotation while not in motion. Screw the wheel to the holes you drilled in the frame earlier. Adjust the height of the wheel so that it fits properly relative to your chosen front wheels. We used 2cm screws, as shown in Figure 8-43. Figure 8-44 shows the robot with its front and rear wheels attached.

Figure 8-41. *Servo horn screw*

> *You might want to make the rear a bit higher than the front. When two Soccer Robots are used, this will prevent the front forks on one robot from hitting the Arduino on the other robot.*

Figure 8-42. *Rear wheel*

Figure 8-43. *Rear wheel attached*

Figure 8-44. *Front wheels attached*

Building the Kicker Leg

Build the kicker leg by connecting a servo horn with a piece of metal that's suitable for kicking. We've utilized parts from a disassembled hard drive reading head for this, as shown in Figure 8-45. However, you can also find many suitable parts in a typewriter.

It is important to make a kicker leg that is strong and long enough to reach the ball, but not so long that it hits the ground. Once you have a suitable piece of metal, you're ready to proceed.

Figure 8-45. *Kicker leg with servo horn attached*

Cut a four-arm servo horn so that only one and a half arms remain, forming a right angle. You can see the modified horn on top of the kicker in Figure 8-45. Following the marks in Figure 8-45, use the 3mm bit to drill three screw holes into the arms. Place the servo horn on top of the metal kicker piece and mark the hole positions on it, including the servo's center hole. Drill holes for 3mm screws, and screw the part to the servo horn. Figures 8-46, 8-47, and 8-48 show the servo and kicker leg from various angles.

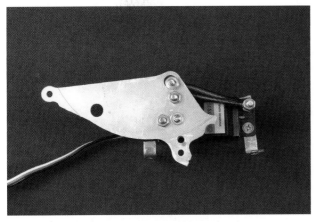

Figure 8-46. *Kicker leg attached to the servo*

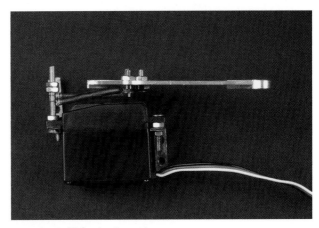

Figure 8-47. *Kicker leg from above*

We attached two springs from an old typewriter to the uppermost screw to ease the servo's movement in the other direction and make the kick punchier. If you have two springs handy, you can attach them, but the mechanism will work fine without them.

Screw the kicking mechanism to the frame, as shown in Figure 8-49.

Figure 8-48. *Kicker leg from the side*

Figure 8-49. *Kicking mechanism*

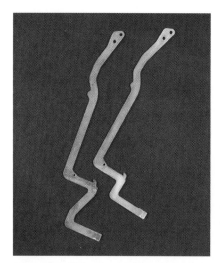

Figure 8-50. *Parts removed from a typewriter*

Adding the Front Fork

The purpose of the front fork is to catch the ball and hold it in place for kicking. We used strips of metal removed from an old typewriter (Figure 8-50), but as long as you can find or cut some strips like the ones shown in Figure 8-52, you'll be fine.

Cut two 9cm parts from the metal strips, as shown in Figure 8-51. Drill screw holes in the positions shown in Figure 8-52 and attach the fork to the frame. Figure 8-53 shows the finished and attached fork.

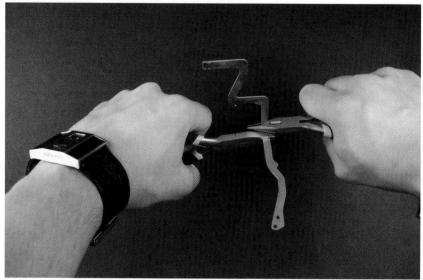

Figure 8-51. *Cutting the front fork*

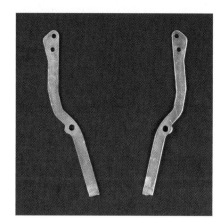

Figure 8-52. *Positions for the screws*

Figure 8-53. *Front fork attached in place*

Attaching the Servos

In total, nine wires come out of three servos. If you try to connect the servos to the Arduino or a prototyping breadboard with jumper wires, problems are likely to arise. This many wires in a moving robot are likely to come loose repeatedly, and repairing them will become pure misery. Because the Arduino is somewhat sensitive, loose wires can cause a short and break it. To solve this problem, you can reduce the number of loose wires by using servo extension cables and an AC lighting connector block. RadioShack sells these blocks under the name "European-style terminal strips."

First, cut a piece from the connector block for four wire holes, as shown in Figure 8-54. Screw red and black wires to it (Figure 8-55).

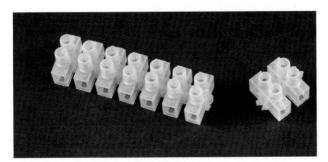

Figure 8-54. *Cutting a suitable connector block*

Figure 8-55. *Wires in place*

Next, you'll connect all the red and black wires of the servos, making these servo connections using three servo extension cables. This way, the servo's own wires do not have to be cut.

For each servo extension cable, cut off the end that does not join to the servo. Solder jumper wires of matching colors to the cut end, as shown in Figure 8-56 (if you need to match a different color to the servo's signal wires, that's OK, but be sure to match red to red and black to black).

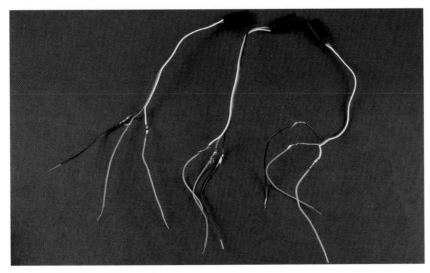

Figure 8-56. *Soldered jumper wires*

To avoid a short circuit, add heat-shrink tubing over the soldered areas (Figure 8-57). We also added larger heat-shrink tubing over the first layer to hold the wires in place nicely, as shown in Figure 8-58. Now screw all the black and red wires to the connector block, as shown in Figure 8-59.

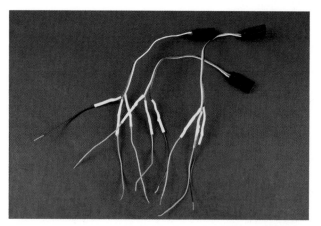

Figure 8-57. *Heat-shrink tubing over the solder joints*

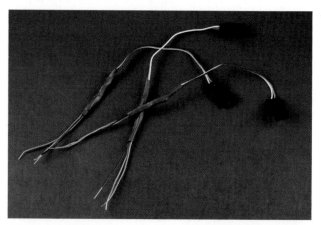

Figure 8-58. *Larger heat-shrink tubing holding the wires in place*

Figure 8-59. *Red and black wires attached*

Putting the Wires, Arduino, and Battery Compartment in Place

Screw the Arduino in place, as shown in Figure 8-60. If you are using the Arduino Pro Mini with a solderless breadboard, you can screw the breadboard in place here.

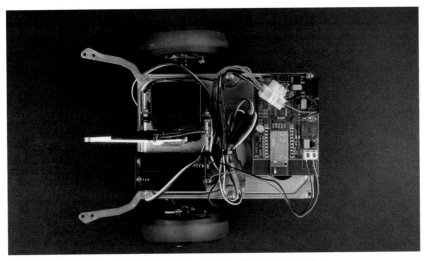

Figure 8-60. *Finished frame of the Soccer Robot, with Arduino in place*

Now attach the connector block wires to the Arduino +5V and GND pins. Connect the servo extensions to the servos and attach the signal wires to the right digital pins. Figure 8-61 shows the wiring diagram.

When you are screwing an Arduino in place, put a felt pad or rubber feet under it to avoid contact with the metallic base; otherwise, you'll short-circuit the Arduino.

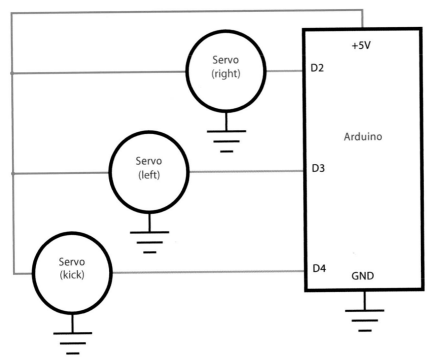

Figure 8-61. *Connection diagram*

Hot-glue the battery compartment to the position shown previously in Figure 8-60. When the battery power cables are attached to the Arduino, the mechanics are complete. Make sure you've connected the power cables correctly (see "Connecting the Arduino to the Bluetooth Mate," earlier in this chapter) so you don't burn out your Arduino.

Programming the Movements

First, program the movements without the cell phone connection or any other factors that could cause things to go wrong. Begin by creating an Arduino sketch for the robot that will enable it to move forward, backward, and turn in both directions while it is stationary.

Moving Forward

This first sketch will move the robot forward, stop it for a moment, and move it forward again.

Write the code in the Arduino development environment and upload it to the Arduino. If you are using the Arduino BT, press the reset button before the transfer. If you are using the Pro Mini, you'll need to detach the Bluetooth Mate and connect a suitable adapter, as described in the "Which Arduino?" sidebar, earlier in this chapter.

Notice that the servos have been attached to the frame so that they point in opposite directions. This sketch needs to rotate them in different directions.

If your robot moves backward, you might have oriented the servos differently (or your servos might turn in a different direction). If that happens, you can modify the sketch to set servoRightPin *to 3 and* servoLeftPin *to 2.*

```
// footballBotForward.pde - Move forward
// (c) Kimmo Karvinen & Tero Karvinen http://BotBook.com

#include <Servo.h>

int servoRightPin = 2; ❶
int servoLeftPin  = 3;
Servo servoRight; ❷
Servo servoLeft;

void moveForward() ❸
{
  servoLeft.write(180); // full speed in one direction
  servoRight.write(0);  // full speed in the other
}

void stopMoving() ❹
{
  servoLeft.write(90);
  servoRight.write(90);
}


void setup() ❺
{
  servoRight.attach(servoRightPin);
  servoLeft.attach(servoLeftPin);
}

void loop() ❻
{
  moveForward();
  delay(1000);
  stopMoving();
  delay(1000);
}
```

Let's have a look at the code:

❶ Define the pins for both servos.

❷ Declare each `Servo` object.

❸ This method sets both servos to full speed in a direction that moves the bot forward.

❹ This method halts both servos.

❺ Initialize each `Servo` object, associating it with its respective Arduino pin number.

❻ Call `moveForward()`, wait for a second after executing it, and call `stopMoving()`. As with all Arduino sketches, the code inside `loop()` executes over and over again, so the robot moves forward for a second, waits, and moves forward again until you switch it off.

Moving in Other Directions

Next, add backward motion as well as right and left turns. Try running this sketch on your bot:

```
// footballBotDirections.pde - Move forward, backward, turn right and left.
// (c) Kimmo Karvinen & Tero Karvinen http://BotBook.com

#include <Servo.h>

int servoRightPin = 2;
int servoLeftPin  = 3;
Servo servoRight;
Servo servoLeft;

void moveForward()
{
  servoLeft.write(180);
  servoRight.write(0);
}

void moveBack() ❶
{
  servoLeft.write(0);
  servoRight.write(180);
}

void turnRight() ❷
{
  servoLeft.write(180);
  servoRight.write(180);
}

void turnLeft() ❸
{
  servoLeft.write(0);
  servoRight.write(0);
}

void stopMoving()
{
  servoLeft.write(90);
  servoRight.write(90);
}
```

```
void setup()
{
  servoRight.attach(servoRightPin);
  servoLeft.attach(servoLeftPin);
}

void loop()
{
    moveForward();
    delay(1000);
    stopMoving();
    delay(1000);

    moveBack();
    delay(1000);
    stopMoving();
    delay(1000);

    turnRight();
    delay(1000);
    stopMoving();
    delay(1000);

    turnLeft();
    delay(1000);
    stopMoving();
    delay(1000);
}
```

This behaves just like the previous example, but with the addition of three new movements (back, right, and left):

❶ When the robot moves backward, the wheels rotate at full speed in the opposite direction from when it moves forward.

❷ To turn the robot right, the left wheel rotates at full speed forward with the right wheel at full speed backward.

❸ To turn the robot left, the right wheel rotates at full speed forward with the left wheel at full speed backward.

Kicking

Here's the code to activate the kicker, which is a standard (as opposed to continuous rotation) servo:

```
// footballBotKick.pde - Kick
// (c) Kimmo Karvinen & Tero Karvinen http://BotBook.com

#include <Servo.h>

int servoKickPin = 4; ❶
Servo kickerServo;

int kickerNeutral = 130; ❷
int kickerKick    = 10; ❸
long kickerWait   = 750; ❹
```

```
void kick()  ❺
{
  kickerServo.write(kickerKick);
  delay(kickerWait);
  kickerServo.write(kickerNeutral);
}

void setup()  ❻
{
  kickerServo.attach(servoKickPin);
  kickerServo.write(kickerNeutral);
}

void loop()  ❼
{
    kick();
    delay(5000);
}
```

Let's look at the code:

❶ The kicker servo is attached to pin 4.

❷ Because you are using a standard servo for the kicker, these values specify a position, not a speed. This position pulls the leg back, ready to kick.

❸ Define how the kick is performed, moving the leg to a specified position. You will probably need to use trial and error to find the right values for this and the neutral position.

❹ This number specifies, in milliseconds, how long to wait for the servo to reach the kick position before returning it to neutral.

❺ This method kicks the servo forward, waits, and then returns it to neutral.

❻ In setup(), attach the Servo object to its pin and set the servo to a neutral position.

❼ Every five seconds, kick!

Before you do anything else, make sure you are happy with the movement of the kicker. It is easier to test and modify the variables at this stage, because this sketch is simple. If you make any changes to these variables, be sure to change them in the subsequent sketches as well.

Controlling Movement from a Computer

Now we'll control the code from the serial console. This way, you will be able to test controlling the robot remotely before controlling it from a cell phone. In the next stages, you can use this capability for your own applications.

The following sketch combines the movement and kick functions from earlier sketches and adds a protocol to control them from the serial console.

```
// footballBotSerialControl.pde - Call move functions from serial console.
// (c) Kimmo Karvinen & Tero Karvinen http://BotBook.com

#include <Servo.h>
```

Do you remember the graphical user interface that we created for controlling the smart home remote control in Chapter 7? By combining those techniques with the ones in this chapter, you could control the Arduino from a computer in style and wirelessly.

If you're using the Arduino Pro Mini (or another Arduino without built-in Bluetooth) with the Bluetooth Mate, don't forget to reconnect the Bluetooth Mate after you've programmed the Pro Mini.

```
int servoRightPin = 2;
int servoLeftPin  = 3;
int servoKickPin = 4;

Servo kickerServo;
Servo servoRight;
Servo servoLeft;

int kickerNeutral = 130;
int kickerKick    = 10;
long kickerWait   = 750;

void kick()
{
  kickerServo.write(kickerKick);
  delay(kickerWait);
  kickerServo.write(kickerNeutral);
}

void moveForward()
{
  servoLeft.write(180);
  servoRight.write(0);
}

void moveBack()
{
  servoLeft.write(0);
  servoRight.write(180);
}

void turnRight()
{
  servoLeft.write(180);
  servoRight.write(180);
}

void turnLeft()
{
  servoLeft.write(0);
  servoRight.write(0);
}

void stopMoving()
{
  servoLeft.write(90);
  servoRight.write(90);
}

void setup()
{
  servoRight.attach(servoRightPin);
  servoLeft.attach(servoLeftPin);

  kickerServo.attach(servoKickPin);
  kickerServo.write(kickerNeutral);

  stopMoving();
```

```
    Serial.begin(115200); ❶
    Serial.println("Football robot. (c) 2008 Karvinen ");
}

void loop()
{
  if ( Serial.available() ) { ❷
    char ch = Serial.read(); ❸

    switch (ch) {
    case '8': ❹
      moveForward();
      delay(250);
      stopMoving();
      break;
    case '2':
      moveBack();
      delay(250);
      stopMoving();
      break;
    case '6':
      turnRight();
      delay(250);
      stopMoving();
      break;
    case '4':
      turnLeft();
      delay(250);
      stopMoving();
      break;
    case '5':
      kick();
      break;
    }
  }
}
```

Because this sketch is a combination of earlier ones, there are very few changes:

❶ Within setup, initialize the serial port and write a welcome message to it. This serial port is used to communicate over Bluetooth.

❷ Check whether there are any characters for us to read over the serial port.

❸ Create a new variable called ch, read a single letter from the serial port, and store it into ch.

❹ If the character is 8, move the robot forward. Similarly, move backward for 2, move right for 6, move left for 4, and kick for 5. If you look at the arrangement of these numbers on a telephone keypad, you'll see that they correspond to cardinal directions.

To send these characters to the robot, use a serial terminal program over Bluetooth, as described in "Testing the Bluetooth Connection," earlier in this chapter.

Steering with an Android Cell Phone

To control the Soccer Robot with Android, you will need a phone running at least Android 2.1 API level 7, with Bluetooth and an accelerometer. At the time of this writing, more than 86% of Android devices in use meet this requirement.

After reading Chapter 6, you should already be familiar with the basics of the Android.

Creating a Simple User Interface

First, create a simple user interface. You'll use the phone's vibrating motor and configure the program to request the privileges it needs to access the motor. At the same time, you'll revisit the graphical user interface programming covered in Chapter 6.

The user interface shows the program's status in one line. In subsequent versions, it will display acceleration readings that change constantly. You'll create a larger text view under it, which will have messages that change less frequently.

Create a new project for the simple interface with the settings shown in Table 8-1 and Figure 8-62. If you need to review the process for creating an Android project, see "Beginning with "Hello World" in Chapter 6.

Table 8-1. Project settings for vibration sample

Setting	Value
Project name	UiAndVibra
Build Target	2.1
Application name	User Interface and Vibra
Package name	fi.sulautetut.android.uiandvibra
Create Activity	UiAndVibra
Min SDK Version	7

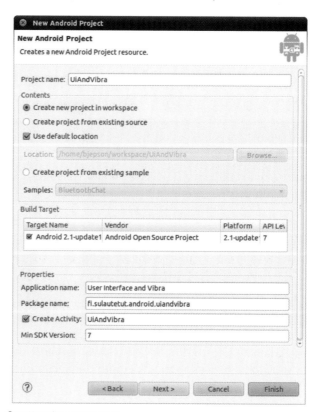

Figure 8-62. *Creating the new project*

Replace the contents of *UIAndVibra.java* (it's under the *src/fi.sulautetut.android* *.uiandvibra* folder on the left side of the Eclipse window) with the following code:

```java
// (c) Tero Karvinen & Kimmo Karvinen http://BotBook.com

package fi.sulautetut.android.uiandvibra;

import android.app.Activity; ❶
import android.content.Context;
import android.content.pm.ActivityInfo;
import android.os.Bundle;
import android.os.Vibrator;
import android.view.WindowManager;
import android.widget.LinearLayout;
import android.widget.TextView;

public class UiAndVibra extends Activity {
    TextView statusTv;
    TextView messagesTv;

    @Override
    public void onCreate(Bundle savedInstanceState) {
        super.onCreate(savedInstanceState);
        initGUI(); ❷
    }

    @Override
    public void onResume() { ❸
        super.onResume();
        statusTv.setText("One-line status. "); ❹
        messagesTv.append(
          "This message box \nwill have\n many lines of text... ");
        vibrate();
    }

    void initGUI() ❺
    {
        // Window
        setRequestedOrientation(
                ActivityInfo.SCREEN_ORIENTATION_LANDSCAPE);
        getWindow().setFlags(
                WindowManager.LayoutParams.FLAG_KEEP_SCREEN_ON,
                WindowManager.LayoutParams.FLAG_KEEP_SCREEN_ON);
        // Contents
        LinearLayout container=new LinearLayout(this);
        container.setOrientation(android.widget.LinearLayout.VERTICAL);
        statusTv = new TextView(this);
        container.addView(statusTv);
        messagesTv = new TextView(this);
        container.addView(messagesTv);
        // Show
        setContentView(container);
    }

    void vibrate()
    {
        Vibrator vibra = (Vibrator) getSystemService(
                Context.VIBRATOR_SERVICE); ❻
        vibra.vibrate(200); ❼
```

```
        }
    }
```

For the most part, the program contains code similar to the examples in Chapter 6. Let's take a look (but don't run the program just yet):

❶   Bring in all the necessary classes.

❷   The `onCreate()` method is a natural place to create a user interface. The actual work of creating the user interface occurs within the `initGUI()` method.

❸   This method is called automatically when the program is ready to interact with the user. This method also runs when you navigate back to it after it's been in the background. Test it by clicking the cell phone's home button and then returning to the program. You'll be rewarded with a little vibration, because the `onResume()` method calls `vibrate()`.

❹   You can write over the `textView` using the `setText()` method, or add more text to the `textView` using the `append()` method. Text views do not scroll automatically, and a later version of the program will erase a text view that is filling up with text.

❺   Create the whole interface programmatically, just like the examples in Chapter 6. If necessary, lock the screen position, keep the display on, and create views with the examples from Chapter 6.

❻   This line enables vibration, but as you'll see in a moment, it requires the correct permissions to succeed. Other services offered by the `get SystemService()` include `LOCATION_SERVICE` (GPS) and the wireless local area network control management `WIFI_SERVICE`. Because `get SystemService()` returns a more generic type, we need to cast it to `Vibrator` with a typecast: `(Vibrator)`.

❼   Make the phone vibrate by calling the `vibra` object's `vibrate()` method, which takes the length of vibration in milliseconds as its parameter. In this example, the phone vibrates for 0.2 seconds.

Adding Permission to Vibrate

Try to run the program on the phone. Does it vibrate?

Instead of vibrating, the program ends (see Figure 8-63), because it tried to use a feature that it does not have permission to use. Permissions are required for certain hardware capabilities for several reasons: vibration can be annoying to users, Bluetooth allows other devices to connect, and the GPS device has privacy implications.

You'll need to add permission to vibrate. On the left side of the Eclipse window, double-click your project manifest: AndroidManifest.xml.

The manifest view appears. At the bottom of the view, click the tab labeled AndroidManifest.xml. Add the following permission just after the `<uses-sdk>` element, as shown in Figure 8-64:

```
<uses-permission android:name="android.permission.VIBRATE" />
```

Figure 8-63. *The program stopping because vibration is not allowed*

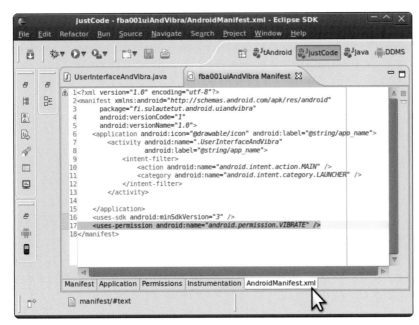

Figure 8-64. *Authorizing vibration using the manifest*

When you run the program again, the phone should vibrate.

The program stays visible, and the user will see the text displayed on the screen (Figure 8-65). Because the program does not react to any other events, nothing else will happen.

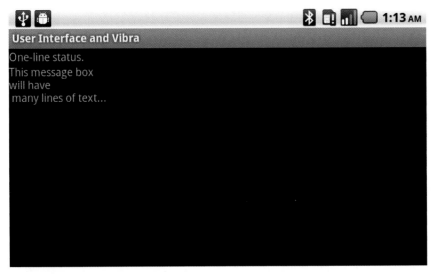

Figure 8-65. *The simple user interface*

The Accelerometer

The *accelerometer* is a sensor that can monitor the cell phone's movements. Gravity causes a downward acceleration, and the accelerometer uses this to determine the position of the cell phone.

The accelerometer reports acceleration relative to three axes: x, y, and z. The largest value (1g, or one unit of gravitational acceleration) for each axis from gravity is:

y (on a table)
> Cell phone on the table, display pointing up

x (held vertically)
> Cell phone held in portrait mode, home and menu buttons on the bottom

z (held horizontally)
> Cell phone held in landscape mode, buttons on the right

Create a new project with the settings shown in Table 8-2 and Figure 8-66. This project will display the accelerometer readings.

> *Freefall acceleration g on the earth is 9.81 meters per second squared. Therefore, the speed of a falling object increases 10 meters per second for each second it falls. That is over 35 km/h per second!*

Table 8-2. *Acceleration project settings*

Setting	Value
Project name	Acceleration
Build Target	2.1
Application name	Acceleration
Package name	fi.sulautetut.android.acceleration
Create Activity	Acceleration
Min SDK Version	7

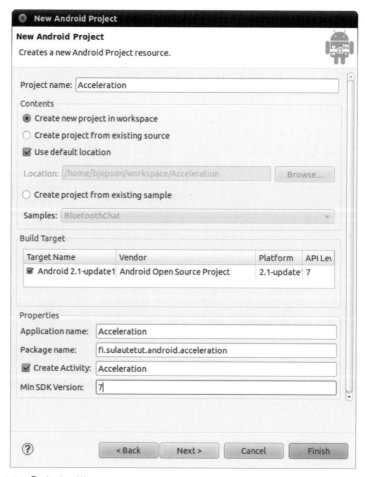

Figure 8-66. *Project settings*

Replace the contents of *Acceleration.java* with the following code:

```java
// Acceleration - print accelerometer values
// (c) Tero Karvinen & Kimmo Karvinen http://BotBook.com

package fi.sulautetut.android.acceleration;

import android.app.Activity;
import android.content.pm.ActivityInfo;
import android.hardware.Sensor;
import android.hardware.SensorEvent;
import android.hardware.SensorEventListener;
import android.hardware.SensorManager;
import android.os.Bundle;
import android.view.WindowManager;
import android.widget.LinearLayout;
import android.widget.TextView;

public class Acceleration
    extends Activity
    implements SensorEventListener  ❶
{
    TextView statusTv;  ❷
    TextView messagesTv;
    SensorManager sensorManager;
    Sensor sensor;
    float g=9.81f; // m/s**2  ❸
    float x, y, z; // gravity along axis, times earth gravity  ❹

    /*** Main - automatically called methods ***/

    @Override
    public void onCreate(Bundle savedInstanceState) {
        super.onCreate(savedInstanceState);
        initGUI();
    }

    @Override
    public void onResume()
    {
        super.onResume();
        initAccel();  ❺
        msg("Running. ");
    }

    @Override
    public void onPause() {
        super.onPause();
        closeAccel();  ❻
        msg("Paused. \n");
    }

    @Override
    public void onSensorChanged(SensorEvent event) {  ❼
        x=event.values[1]/g;    // earth gravity along axis results 1.0
        y=event.values[2]/g;
        z=event.values[0]/g;
        statusTv.setText(String.format(  ❽
                "x: %3.2f y: %3.2f, z: %3.2f",
                x, y, z));
```

```
        }

        @Override
        public void onAccuracyChanged(Sensor sensor, int accuracy) {
            // Must have when Activity implements SensorEventListener.
        }



        /*** Accelerometer ***/

        void initAccel()
        {
            msg("Accelerometer initialization... ");
            sensorManager=(SensorManager) getSystemService(SENSOR_SERVICE); ❾
            sensor=sensorManager.getDefaultSensor( ❿
                    Sensor.TYPE_ACCELEROMETER);
            sensorManager.registerListener(
                    this, ⓫
                    sensor, ⓬
                    sensorManager.SENSOR_DELAY_NORMAL ⓭);
        }

        void closeAccel()
        {
            msg("Accelerometer closing... ");
            sensorManager.unregisterListener(this, sensor);
        }



        /*** User interface ***/

        void initGUI()
        {
            // Window
            setRequestedOrientation(
                    ActivityInfo.SCREEN_ORIENTATION_LANDSCAPE);
            getWindow().setFlags(
                    WindowManager.LayoutParams.FLAG_KEEP_SCREEN_ON,
                    WindowManager.LayoutParams.FLAG_KEEP_SCREEN_ON);
            // Contents
            LinearLayout container=new LinearLayout(this);
            container.setOrientation(android.widget.LinearLayout.VERTICAL);
            statusTv = new TextView(this);
            container.addView(statusTv);
            messagesTv = new TextView(this);
            messagesTv.setText("");
            container.addView(messagesTv);
            setContentView(container);
            msg("User interface created. ");
        }

        public void msg(String s)
        {
            if (7<=messagesTv.getLineCount()) messagesTv.setText(""); ⓮
            messagesTv.append(s);
        }
    }
```

Run the program in the cell phone. When you tilt the phone, you will see the accelerometer readings on the top.

The readings are in relation to the earth's gravity, so approximately 1.0 is the maximum of each axis when the phone is stationary. But what is the largest number you can shake out of the system?

For the most part, the program consists of the same elements as the previous user interface example:

❶ The class must declare that it is listening to the sensors in order to enable any functions related to the accelerometer.

❷ Declare several attributes (variables visible to the entire object). Sensor Manager controls all sensors, including gyroscope, lighting, magnetic field, compass direction, pressure, proximity, and heat. Later, the variable sensor will store the object that controls the accelerometer.

❸ Specify units in the comments because they are not evident within the code. Here, the unit of the variable g is m/s². Gravity affects a steady phone in such a way that it will report acceleration. Acceleration caused by gravity is the same as the freefall acceleration g, which is 9.81 m/s².

❹ Create attributes x, y, and z, in which to store the acceleration relative to each axis.

❺ Initialize the accelerometer in the onResume() method. It is natural to shut off the accelerometer in the onPause() method, which is triggered when the user moves away from the program (for example, by pressing the home button). When the user returns to the program, onResume() is triggered and the accelerometer initializes again.

❻ When the user navigates away from the program, the onPause() method runs. Then the accelerometer closes.

❼ In initAccel, the accelerometer is configured to repeatedly send events to this class, which results in it invoking onSensorChanged(). Each event includes the most up-to-date data. Here, an event is an array that specifies the acceleration relative to each axis (x, y, z). The program stores these values to the object's attributes x, y, and z. This way, the current readings are visible to all our methods.

❽ Display the x, y, and z values. The program could put them in the text views, but it is hard to read unformatted numbers like x=–0.08191646 y=1.0107658 z=0.031933535. Luckily, it is possible to format text in Java, just like in C or C++, using *printf*-style format strings. For example, the first %3.2f is a placeholder for the first variable (x) that follows. 3 is the number of digits before the decimal point, 2 is the number of digits after the decimal pointnd f indicates that the placeholder expects a floating-point value.

❾ Call for the SensorManager class object using the getSystemService() method and store it to the sensorManager variable. Notice the upper-case first letter for the class (SensorManager) and the lowercase first letter for the object (sensorManager). The object returned by the

If Eclipse reports the error "The method onAccuracyChanged(Sensor, int) of type Acceleration must over-ride a superclass method," you'll need to make a change to your project. Click Window→Preferences (Eclipse→Preferences on the Mac), then choose Java→Compiler→Configure Project Specific Settings, and set the Acceleration project's "Compiler compliance level" to 1.6.

getSystemService() function is cast to the right type with (SensorManager).

⑩ Store the accelerometer in the sensor class variable.

⑪ At the end of the initialization, register the class to receive events from the sensor. The parameter of the call this refers to the object where the method call is located (in this case, an instance of our Acceleration class). At the beginning of the class, the program declared that it would implement the SensorEventListener in reference to the class name, which makes it possible for this class to listen to these events.

⑫ The sensor to be listened to is the object just acquired: sensor.

⑬ Define how the program will receive updates. The Soccer Robot reacts much slower than the accelerometer can feed data, so you don't want to get data at full speed. Refresh rates from the fastest to slowest are: SENSOR_DELAY_FASTEST, SENSOR_DELAY_GAME, SENSOR_DELAY_UI, and SENSOR_DELAY_NORMAL. This example uses the slowest refresh rate: SENSOR_DELAY_NORMAL.

⑭ The user interface has one minor improvement: if the screen gets full, msg() erases the messagesTv text view. Otherwise, the latest texts would disappear from the view when the messagesTv fills.

Now you can use the accelerometer (see Figures 8-67 and 8-68). What other programs could you create using it?

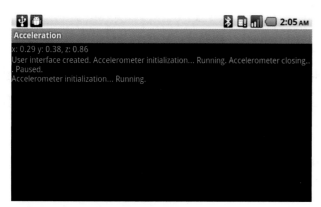

Figure 8-67. *The accelerometer in use*

Figure 8-68. *The accelerometer when the phone is on a table*

An Easier Approach to Bluetooth

Bluetooth can be tricky to work with; it is a complex wireless protocol with lots of features, which means it has many areas where something could go wrong. Therefore, you should mentally prepare for plenty of troubleshooting when you first start using it.

A serial port is one of the most popular ways to transmit data between devices. Bluetooth makes the serial port wireless. In exchange for dealing with Bluetooth's troublesome nature, we get the opportunity to build new types of wireless devices.

We solved some of the more difficult Bluetooth challenges for you by writing an easy-to-use library called tBlue (the appendix covers tBlue in more detail). With this library, you need to use only four commands.

Even though most Android phones promise in their advertisements to support Bluetooth, some older phones don't support the features we use in this example. The Bluetooth API (*application programming interface*) is offered only starting at Android 2.0 API level 5 and in newer phones.

In addition, some manufacturers' Bluetooth implementation is unpredictable. Before you try this next project with your phone, search online for your phone model plus the keywords "Bluetooth SPP" (short for *serial port profile*), as in "bluetooth SPP HTC Desire" (without the quotes).

We tested this project with a Google Nexus One and a Sprint EVO 4G.

Blinking LEDs with Bluetooth

Create a new project with the settings shown in Table 8-3 and Figure 8-69.

Table 8-3. *Bluetooth client project settings*

Setting	Value
Project name	TBlueClient
Build Target	2.1
Application name	Simple tBlue Client
Package name	fi.sulautetut.android. tblueclient
Create Activity	TBlueClient
Min SDK Version	7

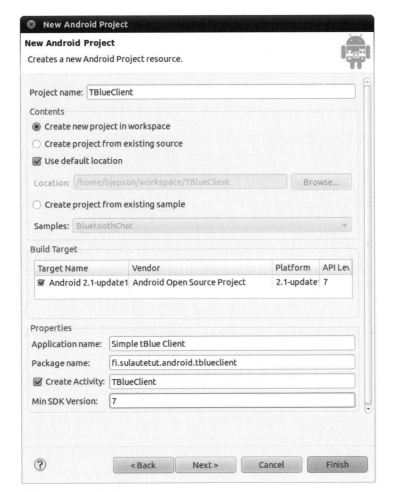

Figure 8-69. *Project settings*

Replace the contents of *TBlueClient.java* with the following code:

```java
// tBlueClient.java - read from serial port over Bluetooth
// (c) Tero Karvinen & Kimmo Karvinen http://BotBook.com

package fi.sulautetut.android.tblueclient;

import android.app.Activity;
import android.os.Bundle;
import android.widget.LinearLayout;
import android.widget.TextView;


public class TBlueClient extends Activity {
    TBlue tBlue;
    TextView messagesTv;

    @Override
    public void onCreate(Bundle savedInstanceState) {
        super.onCreate(savedInstanceState);
        initGUI();
    }

    @Override
    public void onResume() ❶
    {
        super.onResume();
        tBlue=new TBlue("00:07:80:83:AB:6A"); // You must change this! ❷
        if (tBlue.streaming()) {
            messagesTv.append("Connected succesfully! ");
        } else {
            messagesTv.append("Error: Failed to connect. ");
        }
        String s="";
        while (tBlue.streaming() && (s.length()<10) ) { ❸
            s+=tBlue.read(); ❹
        }
        messagesTv.append("Read from Bluetooth: \n"+s);
    }

    @Override
    public void onPause()
    {
        super.onPause();
        tBlue.close(); ❺
    }

    public void initGUI()
    {
        LinearLayout container=new LinearLayout(this);
        messagesTv = new TextView(this);
        container.addView(messagesTv);
        setContentView(container);
    }
}
```

Creating a Bluetooth Connection, Section by Section

In the next step, you'll need to know the Bluetooth address of your Arduino BT or (if you're using another Arduino model) your Bluetooth Mate. If there are any letters (A–F) in the address, type them in uppercase.

The Bluetooth address (in the form XX:XX:XX:XX:XX) might be written on the device, but you can also find it from the computer you paired in "Connecting the Arduino to the Bluetooth Mate," earlier in this chapter:

Windows

> Click the Start menu and choose "Devices and Hardware." Locate the Arduino BT or Bluetooth Mate (listed as a FireFly or RN42 device), and double-click it to bring up the Properties dialog. Click the Bluetooth tab to find the address.

Mac OS X

> Hold down the Option key and click the Bluetooth menu extra. Hover your mouse over the Arduino BT or Bluetooth Mate (listed as a FireFly or RN42 device). Because you are holding down Option, you'll see some extra information, including the Bluetooth address.

Linux

> Select System→Preferences→Bluetooth Manager. You'll see a list of devices along with the Bluetooth address.

Let's take a look at the code:

❶ Open the Bluetooth connection in the onResume() method.

❷ Create a new TBlue class object. On the same line, specify which Bluetooth device address to create a connection for. You *must* replace the hexadecimal string here with the Bluetooth address of your Arduino BT or Bluetooth Mate. Remember, if there are any letters in the address, you must type them in uppercase.

❸ Because there is not always something to be read, you must wait for data to be available. The method streaming() tells the program that it has a working connection.

❹ Read a character over Bluetooth. You can also use the write() method to send something over Bluetooth.

❺ When the user leaves the program, the Bluetooth connection shuts down in order to save the battery.

Did this feel easy? If so, it's because we hid the most complex sections under our wrapper class, TBlue.

Adding tBlue to the Project

In this program, you'll find repeated references to the tBlue.java class. Let's put that class into the project:

1. Select File→New→Class to bring up the New Java Class dialog.

When you are using the tBlue library with your programs, it needs to use the same package name as your project. If the name is written incorrectly, Eclipse will underline it with red inside the TBlue.java source code file. Hover the cursor over the name of the package, as shown in Figure 8-72, and accept the repair suggestion.

Figure 8-70. *Selecting the new class*

2. Give the class the name TBlue (the name is case-sensitive) and set the package to fi.sulautetut.android.tblueclient. Leave everything else at its default. See Figures 8-70 and 8-71.

3. Replace all the code in this class with the TBlue class code. You can get the code from *http://BotBook.com*, and you can read more about it in the appendix.

Figure 8-71. *Creating the new class*

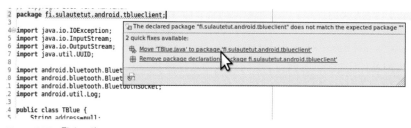

Figure 8-72. *Fixing the error*

Run the program. You will be rewarded with an error message (Figure 8-73) because there are missing permissions.

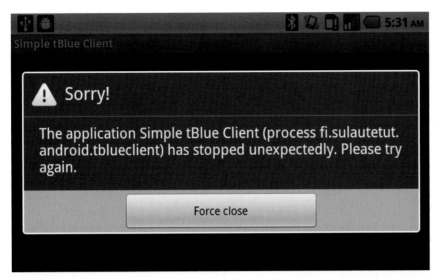

Figure 8-73. *The error message*

Add the missing permissions to the program manifest (Figure 8-74) in the same way as you added the vibration permission in "Adding Permission to Vibrate," earlier in this chapter. These are the permissions required by Bluetooth:

```
<uses-permission android:name="android.permission.BLUETOOTH" />
<uses-permission android:name="android.permission.BLUETOOTH_ADMIN" />
```

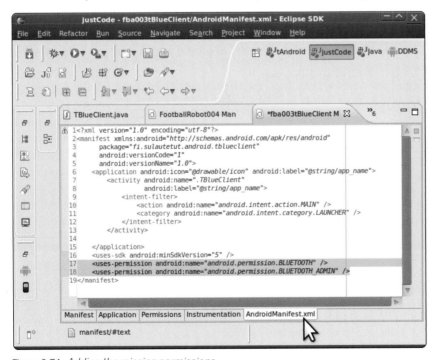

Figure 8-74. *Adding the missing permissions*

Run your program again. The program functions, but no connection was established, because the Arduino is not ready to talk to the Android app. You should just see the error message "Error: Failed to connect," as shown in Figure 8-75.

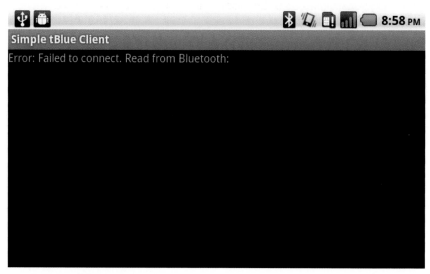

Figure 8-75. *Unable to connect*

Talking to the Arduino over Bluetooth

Open the Arduino development environment and upload the following test program to the Arduino.

```
// serialSample.pde - test serial port over Bluetooth
// (c) Tero Karvinen & Kimmo Karvinen http://BotBook.com

int ledPin =  13;

void setup()
{
  pinMode(ledPin, OUTPUT);
  Serial.begin(115200); // bits per second
}

void loop()
{
  digitalWrite(ledPin, HIGH);
  Serial.println("Hello Serial (over Bluetooth)!");
  delay(500);
  digitalWrite(ledPin, LOW);
  delay(500);
}
```

The program will switch an LED on, display the text, switch off the LED, and wait. This process repeats until you switch it off.

Before you try running your program again, pair your phone with the Arduino BT or the Bluetooth Mate:

If your Arduino does not have an LED connected to pin 13, you will need to connect one now. Connect the LED's long (positive) lead to Arduino pin 13, and connect the LED's short (negative) lead to the GND pin. If your Arduino uses more than 3.3V logic, or an LED that can't tolerate the voltage coming from pin 13, put a resistor in series with one of the LED's leads. The Arduino BT and 3.3V model of the Arduino Pro Mini will use 3.3V on pin 13.

1.   Go to your phone's settings (tap Home, tap Menu, and choose Settings).

2. Tap Wireless & Networks. If Bluetooth is off, turn it on.

3. Tap Bluetooth Settings and tap "Scan for Devices" (make sure the Arduino is powered up).

4. Find your Arduino BT or Bluetooth Mate (FireFly or RN42) in the list. You might need to scroll down the screen to see it. Tap on its entry (it should say "Pair with this device" as well).

5. When prompted for a PIN, use 1234.

Run your program again. If you get a connection error, check the LogCat output (see the appendix). You may also want to try powering up your Arduino just before you run the Android application.

You might need to try opening the Bluetooth connection multiple times from the Android user interface. If the first time is not successful, make the program open the connection again. Press the home key, which automatically runs onPause() and closes the connection. After that, run the program from the Eclipse or directly from the phone, which will try connecting right away.

Finally, the long-awaited words appear on the screen, as shown in Figure 8-76: "Connected successfully! Read from Bluetooth: Hello Serial (over Bluetooth)!" Your Bluetooth connection works.

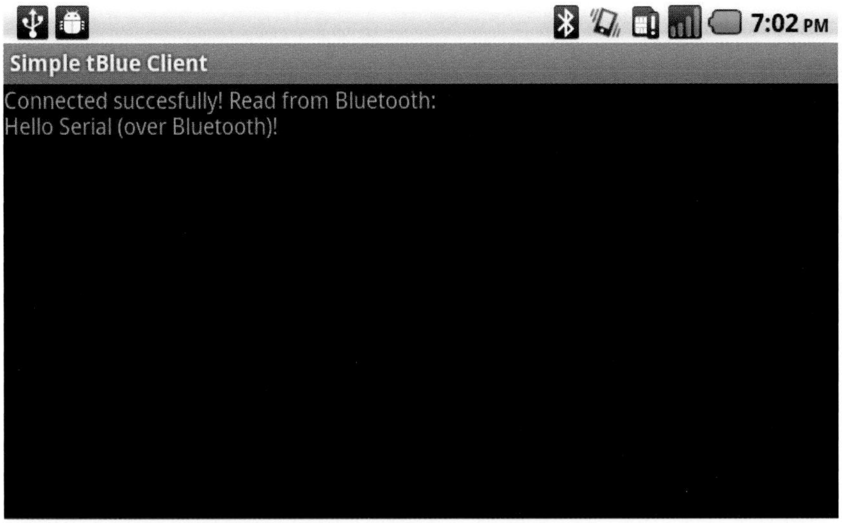

> Make sure that the battery cable does not go over the Arduino BT's or Bluetooth Mate's metal, colored transmitter.

> The final code of the Soccer Robot will make the cell phone retry the connection if it is not successful the first time.

Figure 8-76. *The program working correctly*

Controlling the Robot with Cell Phone Motion

Now it's time to control the Soccer Robot by using the cell phone as a wireless steering wheel. This example combines everything used so far: the accelerometer, serial communications, and Bluetooth.

> *If you are using the 3.3V model of the Arduino Pro Mini, you might encounter problems communicating at 115,200 bits per second. If you have any problems testing out this example, make two changes to your setup.*
>
> *First, connect to your Bluetooth Mate using a serial terminal program, as described earlier in this chapter in "Testing the Bluetooth Connection." Then, type the following commands to enter command mode, change the baud rate to 57.6kbps, and leave command mode (press Enter or Return after each line):*
>
> ```
> $$$
> SU,57.6
> ---
> ```
>
> *Next, power down the Bluetooth Mate (the settings take effect the next time you start it up). Finally, change the following line in the sketch:*
>
> ```
> Serial.begin(115200);
> ```
>
> *to this:*
>
> ```
> Serial.begin(57600);
> ```
>
> *Now the Bluetooth Mate and Arduino will talk at a slightly slower speed.*

Arduino Code

Here's an enhanced version of the Soccer Robot code. Instead of commands that tell the robot which direction to go, the messages now include speed information. For example, the message Sdd-U breaks down like this:

S

> Message start delimiter.

d *(ASCII 100)*

> Move the left wheel at full speed (100%).

d *(ASCII 100)*

> Move the right wheel at full speed (100%).

-

> Don't kick (use k to kick).

U

> Message end delimiter.

Upload the following sketch to your Arduino:

```
// footballrobot.pde - Footballrobot for cellphone control
// (c) Kimmo Karvinen & Tero Karvinen http://BotBook.com

#include <Servo.h>

// Keep track of how far along we are reading a message
const int READY = 1;                  // Ready to receive a message
const int RECEIVED_START = 2;         // Received the start character: 'S'
const int RECEIVED_LEFT_SPEED = 3;    // Received the left speed
const int RECEIVED_RIGHT_SPEED = 4;   // Received the right speed
const int RECEIVED_KICK = 5;          // Received the kick indicator

int state = READY;
```

```
// Define the pins and declare the servo objects.
int servoRightPin=2;
int servoLeftPin=3;
int servoKickPin=4;

Servo kickerServo;
Servo servoRight;
Servo servoLeft;

// Various positions and settings for the kicker.
int kickerNeutral = 130;
int kickerKick    = 10;
long kickerWait   = 750;

// Limit our speed; this needs to be a value between 0 and 90
int maxSpeed = 10;

// Current speeds/kicker setting
int kickNow=0;
int leftSpeed  = 90;
int rightSpeed = 90;

// Temporary speed variables used while we are processing a command.
int newLeftSpeed;
int newRightSpeed;

int ledPin=13;   // LED output pin

void kick()
{
  kickerServo.write(kickerKick);
  delay(kickerWait);
  kickerServo.write(kickerNeutral);
  //Serial.println("Kicking!");
}

void move()
{
  servoLeft.write(leftSpeed);
  servoRight.write(rightSpeed);
}

void stopMoving()
{
  leftSpeed = 90;
  rightSpeed = 90;
}

void setup()
{

//  pinMode(rxPin, INPUT);
//  pinMode(txPin, OUTPUT);
//  mySerial.begin(1200);


  pinMode(ledPin, OUTPUT);
  digitalWrite(ledPin, HIGH);
```

```
            servoRight.attach(servoRightPin);
            servoLeft.attach(servoLeftPin);

            kickerServo.attach(servoKickPin);
            kickerServo.write(kickerNeutral);

            stopMoving();

            Serial.begin(57600);
            digitalWrite(ledPin, LOW);
        }

        void loop()
        {

            if (Serial.available()) {

                int ch = Serial.read(); // Read a character

                switch (state) {
                case READY:
                    if ('S' == ch) {
                        state = RECEIVED_START; // We'll be in this state
                                                // next time through loop()
                    }
                    else if ('?' == ch) {
                        Serial.print("L"); // Let the phone know we are listening
                    }
                    break;

                case RECEIVED_START:
                    if (ch >= 0 && ch <= 10) { // Make sure the values are in range
                        state = RECEIVED_LEFT_SPEED;

                        // Set the temporary left speed value
                        newLeftSpeed = map(int(ch), 0, 10, 90-maxSpeed, 90+maxSpeed);
                    }
                    else { // Invalid input--go back to the ready state
                        state = READY;
                    }
                    break;

                case RECEIVED_LEFT_SPEED:
                    if (ch >= 0 && ch <= 10) { // Make sure the values are in range
                        state = RECEIVED_RIGHT_SPEED;

                        // Set the temporary right speed value
                        newRightSpeed = 180 - map(int(ch), 0, 10, 90-maxSpeed, 90+max-
        Speed);
                    }
                    else { // Invalid input--go back to the ready state
                        state = READY;
                    }
                    break;

                case RECEIVED_RIGHT_SPEED:
                    if ('k' == ch) { // 'k' for kick
                        kickNow = 1;
                    }
                    else { // anything else means don't kick
```

```
      kickNow = 0;
    }
    state = RECEIVED_KICK;
    break;

  case RECEIVED_KICK:
    if ('U' == ch) { // Reached the end of the message

      leftSpeed = newLeftSpeed;    // Set the speeds
      rightSpeed = newRightSpeed;

      if (kickNow) { // Are we supposed to kick now?
        kick();
      }

      // Return to the ready state, clear the kick flag
      state = READY;
      kickNow = 0;
      break;
    }
  }
}

if (state == READY) {
  move();
}
delay(10); // Give the microcontroller a brief rest
}
```

Completing the Soccer Robot

You now have the building blocks for the final project at your disposal: you know how to measure acceleration, you can request whatever permissions your program requires, and you know how to use Bluetooth to communicate between the Arduino and your phone. It's time to put it all together into a Soccer (a.k.a. *football,* in many parts of the world) Robot.

Start a new project for the Soccer/Football Robot using the settings shown in Table 8-4 and Figure 8-77.

Table 8-4. Soccer Robot settings

Setting	Value
Project name	footballRobot
Build Target	2.1
Application name	Football Robot
Package name	fi.sulautetut.android.football
Create Activity	Football
Min SDK Version	7

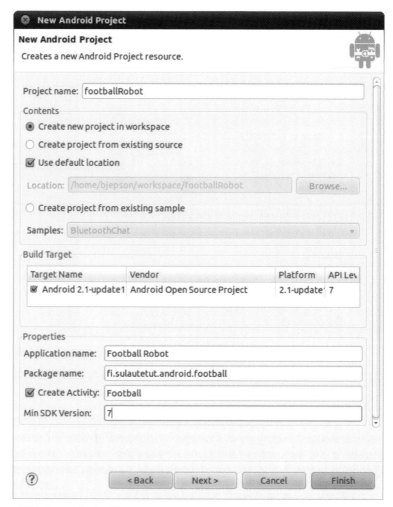

Figure 8-77. *The project settings*

Next:

1. Add the tBlue code to your project, as described earlier in "Adding tBlue to the Project," but use the package name `fi.sulautetut.android.football` instead of `fi.sulautetut.android.tblueclient`. Similarly, change the package name in *TBlue.java* to be the same as that of *Football .java* (`fi.sulautetut.android.football`).

2. Edit your *AndroidManifest.xml* file, as described earlier in "Adding Permission to Vibrate," but add these permissions instead:

   ```
   <uses-permission android:name="android.permission.BLUETOOTH" />
   <uses-permission android:name="android.permission.BLUETOOTH_ADMIN" />
   <uses-permission android:name="android.permission.VIBRATE" />
   ```

3. Here's one more change you'll need to make to *AndroidManifest.xml*. Find these two lines and add the highlighted text:

   ```
   <activity android:name=".Football"
           android:label="@string/app_name" android:configChanges="orientation">
   ```

Finally, replace the contents of *Football.java* with the following example code and set the value of robotBtAddress (highlighted in the following code), as described earlier in "Creating a Bluetooth Connection, Section by Section":

```java
// Football.java - control Arduino over Bluetooth to play ball
// (c) Tero Karvinen & Kimmo Karvinen http://BotBook.com

package fi.sulautetut.android.football;

import android.app.Activity;
import android.content.Context;
import android.content.pm.ActivityInfo;
import android.hardware.Sensor;
import android.hardware.SensorEvent;
import android.hardware.SensorEventListener;
import android.hardware.SensorManager;
import android.os.Bundle;
import android.os.Handler;
import android.os.Vibrator;
import android.util.Log;
import android.view.WindowManager;
import android.widget.LinearLayout;
import android.widget.TextView;

public class Football extends Activity implements SensorEventListener {
    String robotBtAddress="00:07:80:83:AB:6A"; // Change this ❶
    TextView statusTv;
    TextView messagesTv;
    TBlue tBlue;
    SensorManager sensorManager;
    Sensor sensor;
    float g=9.81f; // m/s**2
    float x, y, z, l, r;
    boolean kick;
    int skipped; // continuously skipped sending because robot not ready
    Handler timerHandler; ❷
    Runnable sendToArduino; ❸

    /*** Main - automatically called methods ***/

    @Override
    public void onCreate(Bundle savedInstanceState) {
        super.onCreate(savedInstanceState);
        initGUI(); ❹
        timerHandler = new Handler(); ❺
        sendToArduino = new Runnable() { ❻
            public void run() {
                sendLR(); ❼
                timerHandler.postDelayed(this, 250); ❽
            }
        };
    }

    @Override
    public void onResume()
    {
        super.onResume();
        initAccel();
```

```
            timerHandler.postDelayed(sendToArduino, 1000); ❾

            skipped=9999; // force Bluetooth reconnection ❿
        }

        @Override
        public void onPause() {
            super.onPause();
            r = 0; ⓫
            l = 0;
            sendLR();

            closeAccel();
            closeBluetooth();

            timerHandler.removeCallbacks(sendToArduino); ⓬
            msg("Paused. \n");
        }

        @Override
        public void onSensorChanged(SensorEvent event) {
            x=event.values[1]/g;    // earth gravity along axis results 1.0
            y=event.values[2]/g;
            z=event.values[0]/g;
            updateLR(); ⓭
        }

        @Override
        public void onAccuracyChanged(Sensor sensor, int accuracy) {
            // Must have when Activity implements SensorEventListener.
        }



        /*** User interface ***/

        void initGUI()
        {
            // Window
            setRequestedOrientation(
                    ActivityInfo.SCREEN_ORIENTATION_LANDSCAPE);
            getWindow().setFlags(
                    WindowManager.LayoutParams.FLAG_KEEP_SCREEN_ON,
                    WindowManager.LayoutParams.FLAG_KEEP_SCREEN_ON);
            // Contents
            LinearLayout container=new LinearLayout(this);
            container.setOrientation(android.widget.LinearLayout.VERTICAL);
            statusTv = new TextView(this);
            Log.i("FB", "User interface half way.. ");
            container.addView(statusTv);
            //msg("statusTv added. ");
            messagesTv = new TextView(this);
            messagesTv.setText("");
            container.addView(messagesTv);
            setContentView(container);
        }
```

```
public void msg(String s)
{
    Log.i("FB", s);
    if (7<=messagesTv.getLineCount()) messagesTv.setText("");
    messagesTv.append(s);
}

void vibrate()
{
    Vibrator vibra = (Vibrator) getSystemService(
            Context.VIBRATOR_SERVICE);
    vibra.vibrate(200);
}



/*** Accelerometer ***/

void initAccel()
{
    msg("Accelerometer initialization... ");
    sensorManager=(SensorManager) getSystemService(SENSOR_SERVICE);
    sensor=sensorManager.getDefaultSensor(
            Sensor.TYPE_ACCELEROMETER);
    sensorManager.registerListener(
            this,
            sensor,
            sensorManager.SENSOR_DELAY_NORMAL);
}

void closeAccel()
{
    msg("Accelerometer closing... ");
    sensorManager.unregisterListener(this, sensor);
}



/*** Bluetooth ***/

void initBluetooth()
{
    msg("Bluetooth initialization... ");
    skipped=0; ⓮
    tBlue=new TBlue(robotBtAddress);
    if (tBlue.streaming()) {
        msg("Bluetooth OK. ");
    } else {
        msg("Error: Bluetooth connection failed. ");
    }
}

void closeBluetooth()
{
    msg("Bluetooth closing...");
    tBlue.close();
}
```

```
/*** Motor calculations for left and right ***/

void updateLR()  ⓯
{
    kick=false;
    if (1.5<Math.abs(y)) kick=true;  ⓰
    l=y;
    r=l;
    l+=x;
    r-=x;

    if (l+r<0) { // make reverse turn work like in a car  ⓱
        float tmp=l;
        l=r;
        r=tmp;
    }

    l=constrain(l);
    r=constrain(r);
}

float constrain(float f)  ⓲
{
    if (f<-1) f=-1;
    if (1<f) f=1;
    return f;
}

void sendLR()
{
    if ( (skipped>20) ) {  ⓳
        closeAccel();
        initBluetooth();
        initAccel();
    }
    if (!tBlue.streaming()) {  ⓴
        msg("O");
        skipped++;
        return;
    }

    String s="";  ㉑
    s+="S";
    s+=(char) Math.floor(l*5 + 5);    // 0 <= l <= 10
    s+=(char) Math.floor(r*5 + 5);    // 0 <= l <= 10
    if (kick) s+="k"; else s+="-";
    s+="U";

    statusTv.setText(String.format(
            "%s    left: %3.0f%% right: %3.0f%%, kick: %b.",
            s, Math.floor(l*100), Math.floor(r*100),
            kick));

    tBlue.write("?");  ㉒
    String in=tBlue.read();
    msg(""+in);
    if (in.startsWith("L") && tBlue.streaming()) {
        Log.i("fb", "Clear to send, sending... ");
        tBlue.write(s);  ㉓
        skipped=0;
```

```
    } else {
        Log.i("fb", "Not ready, skipping send. in: \""+ in+"\"");
        skipped++; ㉔
        msg("!");
    }
    if (kick) vibrate(); ㉕
}

}
```

Executing the Program

Let's review the Soccer Robot code in the order of execution. Most of this program is based on the previous examples, so we'll focus only on what's new:

❶ Replace this string with the address of your Bluetooth Mate or Arduino BT. For more information, see "Creating a Bluetooth Connection, Section by Section," earlier in this chapter.

❷ The program creates a task that periodically sends a command to the Arduino. This Handler class manages that task.

❸ This represents the task that sends the command to the Arduino.

❹ When the program launches, it runs onCreate(), where it initializes the app and creates a user interface using our initGUI() method.

❺ Here's where we create the Handler. This is only performed once, at startup.

❻ This line and the ones that follow are a shortcut for creating a new instance of a class and defining its methods (in this case, the run() method), all in one go. This avoids the need to define a new subclass and instantiate it. This case creates a Runnable object (sendToArduino) that invokes sendLR() over and over again. As with all code inside onCreate(), this runs only once, at startup. However, we'll set up a timer inside of onResume() that starts sendToArduino.

❼ Call sendLR().

❽ To have this task run over and over again, use the postDelayed() method and specify a waiting period (250 milliseconds). So, although the accelerometer might call onSensorChanged() much more often, the Arduino is notified only once every 250 milliseconds. Sending messages over Bluetooth can be time-consuming, and if the program were to call sendLR() inside of onSensorChanged(), it would run the risk of introducing a bottleneck into a process that needs to finish quickly. Instead, you're free to let onSensorChanged() react in a timely fashion and can update Arduino using sendLR() at a steady pace.

❾ Set sendToArduino to fire up in 1,000 milliseconds (1 second) and call sendToArduino's run() method, which invokes sendLR() and sets another timer (this time, 250 milliseconds).

❿ The skipped variable tracks how many messages could not be sent to the Arduino. When the program reaches a certain limit—20, which is set in `sendLR()`—the Bluetooth connection resets. So, by setting `skipped` to a ridiculously high number, you can guarantee that the Bluetooth connection will be open the next time it's needed.

⓫ As in earlier examples, the `onPause()` method is invoked when the user navigates away from the app (such as by tapping the home button). First, set `r` and `l` (the right and left speeds) to `0` and call `sendLR()` to stop the robot in its tracks. Also shut down the accelerometer and the Bluetooth connection.

⓬ You don't want to keep talking to the Arduino while the app is paused, so disable `sendToArduino` until `onResume()` is called again.

⓭ Every time `onSensorChanged()` is called, set x, y, and z in the Acceleration project shown in "The Accelerometer," earlier in this chapter. Also call `updateLR()` to convert the accelerometer values to meaningful left/right speeds.

⓮ In `initBluetooth()`, set `skipped` to `0` so that the app knows the Bluetooth connection was retried.

⓯ In `updateLR()`, calculate a suitable speed for the servos from the tilting of the cell phone:

y-axis
> Tilting forward accelerates.

x-axis
> Tilting to the right slows the right servo and speeds up the left one.

z-axis
> Don't take values into consideration.

⓰ Detect `kick` when the phone is shaken enough to send a high value for y.

⓱ For the cell phone steering wheel to function just like in a car, when the robot backs up, the speeds of the wheels must be reversed. When you go forward and tilt the phone right, the right wheel will slow down and the robot will turn right. If you then tilt the phone backward, both wheels will slow down until the speed is negative (i.e., backward). The speed of the right wheel is a smaller value—but because it's negative, the right wheel actually spins faster and the robot turns left. To make the right tilt always turn right, we must flip the left and right wheel speeds when we start moving backward.

⓲ This function makes sure the values for `l` and `r` don't go too high.

⓳ If the `skipped` counter exceeds 20, close and reopen the Bluetooth connection and reinitialize the accelerometer.

⓴ If the `tBlue.streaming()` method returns `false`, it means that the connection is not active. Increment the `skipped` count and do nothing until the next time through this method.

㉑ This string contains the message we're sending to the Arduino. Scale the values of `l` and `r` (which can range from –1 to 1) to range from 0 to 10.

This gives a relatively rough set of values (0–4 for reverse, 5 for stopped, 6–10 for forward) but also means that the controls will not be excessively sensitive, and you won't have to spend as much time holding the phone just right to get the robot to stop.

The Arduino code will scale this to a range of values based on the setting for the maxSpeed variable in the Arduino sketch. In addition, because the servos are in opposite orientations, you'll need to spin them in opposite directions (see "Programming the Movements," earlier in this chapter). So the Arduino sketch uses the scaled value for the rotation speed of one servo, but subtracts that value from 180 to determine the rotation speed of the other one.

㉒ Send a query to Arduino. If the program gets an L in response, it means that Arduino is ready to communicate.

㉓ Send the string to the Arduino. For example, the message Saa-U sends left and right speeds of 97 and indicates that the robot shouldn't kick. However, because the program limits the range of values from 0 to 10, it will send characters in the low end of the ASCII character set, which are generally represented in print with control characters, symbols, or short names. So a speed of 0 would be NUL (0 in ASCII), a speed of 5 would be ENQ (5 in ASCII). The good news is that you don't need to keep track of which ASCII character is which, because the Arduino takes care of converting the characters to values that it can use.

㉔ If Arduino has not sent the ready signal L or the connection is not ready, we skip sending this time and increase the skipped counter. This way, we can reinitialize the connection if we have to skip many messages in a row.

㉕ Vibrate the phone to show the user that his kick has been sent to Arduino.

Playing Soccer

Power up the Soccer Robot and the cell phone. Hold the cell phone in your hand in a horizontal (landscape) position (see Figure 8-78).

Figure 8-78. *Playing soccer via cell phone*

Turn by tilting the cell phone in the air to the right or left. Accelerate by tilting forward, and back up by tilting backward.

When you shake the cell phone sharply, it will vibrate and the robot will kick. Enjoy the game: you deserve it!

What's Next?

You can now congratulate yourself and shake your own hand. You have built all the projects in this book. What combinations of Arduino, motors, Android, Processing, Python, and more can you dream up? Visit us at *http://BotBook .com* and let us know!

tBlue Library for Android

In the Bluetooth examples you've seen in this book, the interactions have been simple: opening a connection and sending a few characters. But to use Bluetooth, you must also:

- Find the cell phone's Bluetooth adapter.

- Define a Bluetooth object that represents the device you are talking to (such as an Arduino).

- Open a socket that represents the Bluetooth serial connection.

- Open some stream objects: one for sending messages, one for receiving.

Because the Android Bluetooth APIs are complex, we created the class TBlue, which keeps things simple. Chapter 8 has example code that shows you how to use the tBlue library. Here is the source of the TBlue class:

```
// tBlue.java - simple wrapper for Android Bluetooth libraries
// (c) Tero Karvinen & Kimmo Karvinen http://terokarvinen.com/tblue

    package fi.sulautetut.android.tblueclient;

    import java.io.IOException;
    import java.io.InputStream;
    import java.io.OutputStream;
    import java.lang.reflect.InvocationTargetException;
    import java.lang.reflect.Method;

    import android.bluetooth.BluetoothAdapter;
    import android.bluetooth.BluetoothDevice;
    import android.bluetooth.BluetoothSocket;
    import android.util.Log;

    public class TBlue {
      String address=null;
      String TAG="tBlue";
      BluetoothAdapter localAdapter=null;
      BluetoothDevice remoteDevice=null;
      BluetoothSocket socket=null;
      public OutputStream outStream = null;
      public InputStream inStream=null;
      boolean failed=false;

      public TBlue(String address)
```

```
{
    this.address=address.toUpperCase(); ❶
    localAdapter = BluetoothAdapter.getDefaultAdapter(); ❷
    if ((localAdapter!=null) && localAdapter.isEnabled()) {
        Log.i(TAG, "Bluetooth adapter found and enabled on phone. ");
    } else {
        Log.e(TAG, "Bluetooth adapter NOT FOUND or NOT ENABLED!");
        return;
    }
    connect(); ❸
}

public void connect()
{
    Log.i(TAG, "Bluetooth connecting to "+address+"...");
    try    {
        remoteDevice = localAdapter.getRemoteDevice(address); ❹
    } catch (IllegalArgumentException e) {
        Log.e(TAG, "Failed to get remote device with MAC address."
                +"Wrong format? MAC address must be upper case. ",
                e);
        return;
    }

    Log.i(TAG, "Creating RFCOMM socket...");
    try {
        Method m = remoteDevice.getClass().getMethod
                ("createRfcommSocket", new Class[] { int.class });
        socket = (BluetoothSocket) m.invoke(remoteDevice, 1); ❺
        Log.i(TAG, "RFCOMM socket created.");
    } catch (NoSuchMethodException e) {
        Log.i(TAG, "Could not invoke createRfcommSocket.");
        e.printStackTrace();
    } catch (IllegalArgumentException e) {
        Log.i(TAG, "Bad argument with createRfcommSocket.");
        e.printStackTrace();
    } catch (IllegalAccessException e) {
        Log.i(TAG, "Illegal access with createRfcommSocket.");
        e.printStackTrace();
    } catch (InvocationTargetException e) {
        Log.i(TAG, "Invocation target exception: createRfcommSocket.");
        e.printStackTrace();
    }
    Log.i(TAG, "Got socket for device "+socket.getRemoteDevice());
    localAdapter.cancelDiscovery(); ❻

    Log.i(TAG, "Connecting socket...");
    try {
        socket.connect(); ❼
        Log.i(TAG, "Socket connected.");
    } catch (IOException e) {
        try {
            Log.e(TAG, "Failed to connect socket. ", e);
            socket.close();
            Log.e(TAG, "Socket closed because of an error. ", e);
        } catch (IOException eb) {
            Log.e(TAG, "Also failed to close socket. ", eb);
        }
        return;
    }
```

```java
        try {
            outStream = socket.getOutputStream(); ❽
            Log.i(TAG, "Output stream open.");
            inStream = socket.getInputStream();
            Log.i(TAG, "Input stream open.");
        } catch (IOException e) {
            Log.e(TAG, "Failed to create output stream.", e);
        }
        return;
    }

    public void write(String s) ❾
    {
        Log.i(TAG, "Sending \""+s+"\"... ");
        byte[] outBuffer= s.getBytes(); ❿
        try {
            outStream.write(outBuffer);
        } catch (IOException e) {
            Log.e(TAG, "Write failed.", e);
        }

    }

    public boolean streaming() ⓫
    {
        return ( (inStream!=null) && (outStream!=null) );
    }

    public String read() ⓬
    {
        if (!streaming()) return ""; ⓭
        String inStr="";
        try {
            if (0<inStream.available()) {
                byte[] inBuffer = new byte[1024];
                int bytesRead = inStream.read(inBuffer);
                inStr = new String(inBuffer, "ASCII");
                inStr=inStr.substring(0, bytesRead); ⓮
                Log.i(TAG, "byteCount: "+bytesRead+ ", inStr: "+inStr);
            }
        } catch (IOException e) {
            Log.e(TAG, "Read failed", e);
        }
        return inStr;
    }

    public void close()
    {
        Log.i(TAG, "Bluetooth closing... ");
        try    {
            socket.close(); ⓯
            Log.i(TAG, "BT closed");
        } catch (IOException e2) {
            Log.e(TAG, "Failed to close socket. ", e2);
        }
    }
}
```

When you're debugging your code, it is handy to see execution details that the end user would find distracting. The log is the right place to display this information because it does not clutter up the user interface. The Android log is handled through a facility known as *LogCat*. To send a message to the log, you can use a line such as the following (replace *myProgram* with something that identifies your application, because many other parts of Android display LogCat messages):

```
Log.e("myProgram", "Your error message here.");
```

You can make LogCat visible in Eclipse by going to Window→Show View→Other→Android→LogCat. If nothing is displayed in the log, choose your phone's device view with Window→Show View→Other→Android→Devices.

When things go wrong, the libraries used with Bluetooth will throw an exception. If you do not catch the exception with a try-catch structure, the program will crash when the exception is thrown. Here's how to catch it and use Log() to log the error:

```
try {
   outStream.write(outBuffer);
} catch (IOException e) {
   Log.e(TAG, "Write failed.", e);
}
```

In the preceding example, we tried to write to a stream named outStream. If that fails, we display the error in the log. Even though we don't do too much to handle the *causes* of the exception, catching it will ensure that the program will not crash, and logging it will help us diagnose the failure.

You'll see many instances where LogCat is used in tBlue.

As you saw in Chapter 8, the user (in this case, you) kicks things off by creating a new TBlue class object with something like the following (replacing the address in parentheses with the Bluetooth address of the Arduino BT or Bluetooth Mate):

```
tBlue=new TBlue("00:07:80:83:AB:6A");
```

Now, let's have a look at the most important parts of the tBlue library:

❶ In the constructor, we store the specified Bluetooth device address to our class's address attribute. Note that Android libraries require letters in Bluetooth addresses to be typed in uppercase.

❷ We will ask the system for an object that can be used for controlling the cell phone's Bluetooth adapter. We store it to the variable localAdapter.

❸ At the end of the constructor, we will call our connect() method.

❹ First, we ask the local Bluetooth adapter to fetch a representation of the remote device (the Arduino BT or the Bluetooth Mate), and then we store it to the remoteDevice attribute.

❺ We now create a new socket using the RFCOMM Bluetooth serial port protocol. Nothing is sent through the socket yet, so this should work even if no Bluetooth device exists yet.

❻ Discovering all possible Bluetooth devices is processor-intensive, so we will interrupt the search once we find the device we are looking for.

❼ Here, we connect the socket. This is this point where errors usually occur if a connection cannot be established to the Bluetooth device.

❽ Finally, we will open a stream for output and input so we can talk and listen to the device.

⑨ Once the connection is open, your code (for example, the Soccer Robot program in Chapter 8) can send messages to the Arduino with a command such as `tBlue.write("A")`.

⑩ Since a regular `String` cannot be used when writing to an output stream, we must convert the text to an array of bytes. Hiding these types of details within `TBlue` makes the code in Chapter 8 easier to read.

⑪ This method will tell your code whether the connection is still active.

⑫ Just as with sending messages, your code can receive a message from the Arduino with a command like `String s=tBlue.read()`.

⑬ Before we do anything, we will check that the connection is open. If not, we just return an empty string. This will save us from getting unnecessary error messages and having to keep checking the results of this command (however, as you can see in Chapter 8, your code should periodically check the state of the connection to make sure it's ready).

⑭ To ensure that we don't include any extraneous characters in the result, we trim `inStr` to have the length specified in `bytesRead`, which we got back from the call to `inStream.read()`.

⑮ We close the connection when exiting the program.

Now you've seen the Android Bluetooth API in more detail than you saw in Chapter 8. The tBlue library lets you forget the complexities of Android's Bluetooth APIs and concentrate on what you want to do in your program.

Index

Symbols

Colophon

The heading and cover font are BentonSans, the text font is Myriad Pro, and the code font is TheSansMonoCondensed. *Make: Arduino Bots and Gadgets* was composed in Adobe InDesign CS4 by Newgen.

About the Authors

Kimmo Karvinen works as CTO of a hardware manufacturer that specializes in smart building technology. Before that, he worked as a marketing communications project leader and as a creative director and partner in an advertising agency. Kimmo's education includes a Masters of Art.

Tero Karvinen teaches Linux and embedded systems at the Haaga-Helia University of Applied Sciences, where his work has also included curriculum development and research in wireless networking. He previously worked as the CEO of an advertising agency. Tero's education includes a Masters of Science in Economics. *www.TeroKarvinen.com*.